SOLAR ENERGY PLANNING

SOLAR ENERGY PLANNING

a guide to residential settlement

PHILLIP TABB

Phillip Tabb Architects, Boulder, Colorado
College of Design and Planning, University of Colorado,
Boulder, Colorado

McGRAW-HILL BOOK COMPANY

New York St. Louis San Francisco Auckland Bogotá
Hamburg Johannesburg London Madrid Mexico
Montreal New Delhi Panama Paris São Paulo
Singapore Sydney Tokyo Toronto

Library of Congress Cataloging in Publication Data

Tabb, Phillip.
  Solar energy planning.

  1. Solar energy.   2. City planning.   I. Title.
TJ810.T27 1984   711'.4   83-24810
ISBN 0-07-062688-X

1234567890 HAL/HAL 8987654

ISBN 0-07-062688-X

The editors for this book were Joan Zseleczky and Susan Killikelly,
the designer was Naomi Auerbach, and the production supervisor
was Reiko F. Okamura. It was set in Auriga by University Graphics, Inc.

Printed and bound by Halliday Lithograph.

To the memory of my grandmother
GEORGINA TABB (1887–1982)

CONTENTS

PREFACE

Since the Arab oil embargo in the early 1970s, energy has become an important force in the design of buildings. In the United States alone, hundreds of thousands of buildings have been constructed using various energy-conservation techniques and solar energy technologies for space or water heating. By the year 1990 there could be as many as 5 million energy-responsive buildings. With this magnitude of growth, the need for solar energy planning is evident, particularly in the residential sector, where most of the solar energy systems will continue to be demonstrated.

In the search to satisfy the need for energy and the many other needs connected to housing, several questions arise. What kinds of solar energy systems are appropriate to the variety of residential building types? What are the energy implications of the use of these systems in multiple buildings? What limitations does density place upon energy-oriented design? In what ways can energy-conscious planning affect contemporary residential settlement patterns? It is the purpose of this book to identify the types of solar technologies presently in use and to describe their potential and limitations for larger-scale utilization.

Solar energy planning concepts ideally should be applied to all scales of the built environment, particularly to most areas of the building sector—residential, commercial, institutional, and industrial—and to the transportation sector. The last several years have seen a shift in interest from residential use of solar energy to commercial applications. Three demonstration cycles of the U.S. Department of Energy have revealed practical solutions to many types of commercial solar architecture. However, most of these demonstrations have been limited to small single buildings. For larger nonresidential buildings, many questions are still unanswered. The predominant use of solar technology remains with residential buildings. For this reason, this book focuses on planning issues at the various levels of *residential* settlement.

The solar energy industry has promoted energy-conservation measures in conjunction with solar mechanical technologies for single buildings. Many of the problems associated with multiple buildings have been overlooked. Protecting solar collection from shadows of adjacent buildings is certainly an important subject for both small and large scales of development. Land use, density, automobile circulation, and open-space design are also issues that can affect the quality of energy conservation. The book is written as a guide for planners and developers, architects and solar engineers, and educators and students so they may be alerted to some of the energy issues and limitations that exist at the larger scales.

The book begins with an introductory chapter of residential development patterns. The influences of energy, especially the use of solar energy in community planning, are described for many cultures and times throughout history. The second chapter presents some limitations of solar energy in an attempt to define realistic boundaries within which creative solar design and planning can occur. The chapter also serves to define state-of-the-art solar energy systems, primarily active and passive systems. The need for solar-access protection and the methods for solar-access planning at varying densities are the subjects of the third chapter. A checklist for solar-access planning is included at the end of the chapter. The first three chapters form a historical and theoretical foundation for the remaining chapters, in which the concepts are applied at various scales.

The remaining four chapters, in sequence, relate solar energy considerations in expanding scales of residential development: individual shelter design, cluster development, neighborhood planning, and residential settlement. The need to plan for solar energy is woven through these various scales from the technical concerns for a thermal zone or room to the broader and more complex concerns for an entirely planned community or new town. Applications of solar energy planning concepts are discussed at each of these scales. At the end of the book are three appendixes. Appendix A presents several simple computational methods for determining shadow patterns and protecting solar access. Appendix B presents information on solar covenants and ordinances. Appendix C is a table of selected conversion factors for use in converting customary measurements to metric, or SI, measurements. Throughout the book, the use of computer-generated graphics helps explain many of the solar planning ideas that otherwise would have been too difficult to illustrate.

The work presented here spans nearly 10 years, although the actual manuscript was prepared in only a year—thanks to my Apple III computer. The ideas and designs presented in this book have evolved from vague visions seeded a decade ago. During that time, many individuals have greatly influenced their growth. First, I would like to give thanks to my students at the University of Colorado who have worked long hours under great pressure to produce some of the work for this book. James Elliott, Stephen Glascock, Deidra Heaton, David Horsley, and Dawn Marine developed many of the case studies illustrated in the book. James Hopperstead, Greg Lemon, Jeff Davis, and Pat Davies produced most of the illustrations. And Scott Wolfe digitized and produced most of the computer-aided drawings found throughout the book.

I would like to acknowledge the assistance and support of Jane Harris, who helped with the research and editing, Nelson Greene, who gave guidance for all of the computer-aided design work, Russ Derickson, who provided solar engineering concepts, Larry Peterson, who reviewed the manuscript, and Jan F. Kreider, who provided hundreds of useful suggestions that greatly improved the technical quality of the book, and my parents, who provided financial support. Very special thanks to my sons, Michael and David, for their aliveness and inspiration and to Myfanwy Lloyd-Tabb for her encouragement and helpful criticism throughout the creation of the manuscript.

Phillip Tabb

INTRODUCTION

A view of the earth from outer space gives our generation a perspective never before experienced in history. It is now difficult to imagine the whole earth in any other way than as a sphere of blue and white surfaces floating in silence around the sun. Given energy from our sun millions of miles away, we are passengers on a planet involved in the intricate cycles of life. The full impact of this new perception is just unfolding; the wonderful changes that will occur as a result of this new picture of the Earth—so truly beautiful and contained—are yet to be manifested.

One way to visualize our life here on earth is to see it from afar. Imagine you are a traveler on a spaceship from Sirius, 25 trillion miles away. Our sun, although average in size, is a raging nuclear reactor 870,000 miles in diameter with nine planets. As you pass through our solar system and view the circling planets, one, with a composition of carbon, oxygen, and hydrogen that supports life, stands out. Approaching this planet and penetrating its atmosphere, you can see two white polar caps with large bodies of water and many land masses beneath the motion and whorls of a white gas. Human settlements on mountains and islands and in river valleys, plains, and deserts are apparent. They vary from small communities to large metropolitan networks. Having traveled so far, you may wonder about the ways in which this civilization relates to its unique solar system and more specifically to this special planet.

What is life on this planet really like? Further investigation would reveal a fairly complex physical and social organization that extends to most regions of this globe. The works of this civilization would be quite evident. There are dams and bridges and great mosaics of farms and fences. Highways stretch from one end of the land masses to the other. There are many airplanes, trains, ships, and automobiles, and there is an occasional space shuttle on the move. Great activity would be apparent. The larger cities stretch for miles, covered by blankets of smoke and other gases. At first glance, it would appear that there is a fairly energetic species at work. Who would suspect that "they"—we—are experiencing an energy crisis when so much energy is seemingly available?

The Sirius solar system and the planet from which you came may be very similar to this one. But perhaps your planet did not experience a fossil fuel era. Your technology, slightly more advanced than earth's, may be based upon star energy, synthetic fuels, and hydrogen. Your cities may be more strongly organized around those energy sources, and your dwellings may also be designed as little energy machines.

This voyage, though just a fantasy, gives us a quick opportunity to see ourselves, the way in which we live, and our physical and technological responses to the limited resources that are available to us. With a critical eye, we can identify those processes that seem to fit into this larger picture, for the next several centuries most certainly will bring greater space travel, more refined uses of our resources, the decline of the fossil fuel era, and new forms of energy.

A view of the earth from outer space gives us a simplified picture—the earth, our moon, and the sun in a dark, expansive context dotted with stars, planets, and other "travelers" in space. Certainly the relationship between the sun and the earth is a powerful one. In fact, most of the earth's energy—including most thermal energy and the energy in all of our food and the fossil fuels we are now consuming—is derived from the sun. It seems clear that we should evolve a stronger relationship to solar energy. This is not an easy task. With 4 billion fellow residents and social, economic, political, and physical structures very much based upon diminishing fossil fuels, we may need several centuries to accomplish the task. Planning now for greater utilization of solar energy can ease the transition from an age of fossil fuels and ensure plentiful energy for future generations.

Solar energy planning is concerned with creating another structure, one that allows for contemporary needs, is sensitive to the past, responds to the limitations of solar energy, and is more integrated with the forces of our universe. Solar energy planning is a design response. Decisions being made today are affecting this potential. The process can be organic; that is to say, it can grow from the prevailing planning and architectural methodologies. It can evolve slowly toward the building of solar buildings and solar communities. The opportunities are worldwide, and the challenge is exciting. A shift in this direction could reduce some of the preoccupation with nationalism, defense, isolation, and survival. We could be made more aware of the global community that we really are: one people, one planet.

Woven throughout solar energy planning is the concept of scales of response. Certain conditions must exist and be preserved for solar energy to be effectively utilized, for photosynthesis to take place, and for solar heating or solar power generation to be possible. Solar energy planning affects design decisions at various scales so that these conditions are encouraged and maintained. Solar energy planning involves many positive limitations on physical design at several scales:

1. Shelters (individual dwellings)
2. Clusters (small groupings of dwellings)
3. Neighborhoods (larger groupings or small communities)
4. Residential settlements (communities and new towns)

Imagine space travelers from Sirius experiencing the earth in 200 years. They might see dwellings with solar electric cells; clusters with shared energy systems; neighborhoods lush with gardens; and communities based upon large-scale solar collection, urban farms, and the use of solar electric vehicles. Perhaps there will be solar-powered space colonies. Many obstacles stand between the built environment and this vision. We must decide whether or not we will plan for solar energy.

A fitting ending to this introduction is a quote by Nigel Calder, author of *Spaceships of the Mind.*

> The human world, past and future, is shaped by constant pressure from the imagination and ambition of individuals who have big ideas. By a big idea I mean one that will prevail not by degree or even by persuasion but because it captures the enthusiasm of people who will struggle against great difficulties to make it happen.*

*N. Calder, *Spaceships of the Mind*, British Broadcasting Corporation, London, 1978.

CHAPTER 7

DEVELOPMENT PATTERNS

*Man is a singular creature. He has a set of gifts
which make him unique among the animals: so
that, unlike them, he is not a figure in the
landscape—he is a shaper of the landscape.*
J. BRONOWSKI*

People historically have bound together to form communities of various kinds. The physical and spiritual manifestations of these settlements have satisfied many needs, but principally their function has been to facilitate each group member's attainment of commodities necessary for life support and to contribute to the realization of individual emotional and intellectual needs and aspirations. The human settlement is a means of celebrating and coping with environmental changes as well as with fellow beings.

Ancient societies, disposed all over the globe, evolved similar city plans with orientation to the cardinal points of the compass so that all of the buildings, placed within a grid, would receive solar energy during the winter months. The Egyptians, the Greeks, the Chinese, the Aztecs, the Incas, and the Anasazi Indians alike created solar communities and architecture in response to the movement of the sun. For centuries these people lived with the energy from the sun and charcoal or wood. As wood and charcoal supplies dwindled, greater reliance was placed upon solar energy, and as a result, access to the sun became a stronger and more important planning determinant.

In modern times, much of this was lost. With vast quantities of coal, oil, and natural gas and with the promise of inexhaustible supplies of

*From Bronowski.[6] Copyright © 1973 by J. Bronowski.

nuclear energy, modern societies evolved city plans of undisciplined sprawl. World population exploded with over 4 billion inhabitants. Cities grew up and out using vast resources. Complex problems arose caused by rapid growth and changing technology. Complicated transportation networks, diverse and problematic land use, complex city services and maintenance schedules, water and waste distribution and purification, pollution, crime and other social problems—all added to the milieu of contemporary residential settlement. The confluence of these phenomena overshadowed direct use of the sun. It was not until recently that fossil fuels were recognized as being finite and nonrenewable; and it was not until actual experiences of energy shortages had occurred that solar energy was seen as a serious alternative.

A look at current energy-oriented design in relation to historical development is presented in this chapter. *Ancient development, post-industrial-revolution development,* and *modern development* are all compared in relation to energy sources and consumption. Contemporary *energy-responsive development* is discussed and illustrated, for the use of solar energy for heat and light is once again becoming an integral part of the planning process.

HISTORICAL DEVELOPMENT

The origin of humankind is believed to have occurred some 5 million years ago near the equator in central Africa. Ancient stories put our species in a golden age, perhaps in the Garden of Eden, which may have been located in this place. Energy was probably not an important issue as the earth was very plentiful. Cold temperatures and hunger were probably rarely experienced. As time went on, people hunted and gathered and then began to migrate to other lands where they most likely found the need for additional food, clothing, housing, and fire for cooking. Ancient camps demonstrated basic environmental responses through location adjacent to rivers or streams, orientation to the sun, and breaks to the prevailing winds. These early camps may have used simple passive solar energy techniques for heating.

As villages grew into communities and communities grew into cities, residential settlements began to evolve development patterns maintaining orientation to the sun for warmth and light. Many ruins still exist today that demonstrate this pattern. Paths and streets were oriented along the east-west axis in order to allow for good solar access for each building. Buildings were designed to accept sunlight during the winter months and to provide shading in the summertime. Socrates, in the fourth century B.C., spoke of these simple design principles: "In houses that look toward the south, the sun penetrates the portico in winter, while in summer the path of the sun is right over our heads above the roof so that there is shade."[8]

Early Responses to the Sun

Early Paleolithic camps, which are 300,000 years old, have been found in southern France around the Mediterranean Sea. Although there is no real physical evidence that the camps were primarily organized in response to the sun, the ruins seem to characterize communal life as it may have occurred at that time. Proximity to food and water, access to solar energy, protection from the winds, and views for defense seem to have been the primary planning forces. The camps were small enough that response to these forces was probably quite natural and uncomplicated. Figure 1-1 is an illustration of a camp that may have developed during this time period. As the camps grew into villages and as life became more diversified, the physical form required more order and planning. Beginning around 2800 B.C., several prehistoric monuments, such as those found at Stonehenge, Avebury, and Carnac, reveal great sensitivity and extraordinary discipline to the movement of the sun.

Stonehenge, located in Wiltshire on Salisbury Plain, England, evolved in four distinct phases and was used as a temple and burial ground for seventeen centuries. The ultimate design was made of concentric circles of stone. Going from the outside in, the circles included an outer earth berm, the Avebury holes and Y-shaped holes, the Sarsen circle of standing stones with curved lintels, the bluestone circle, the five giant triliths, and the altar stone. The five triliths in a horseshoe configuration weigh between 20 and 24 tons each. The altar stone and the heel stone located outside the monument proper were aligned with the mid-summer sunrise. Many other movements of the sun, the moon, and the wandering stars (the five visible planets) were observable within the monument. It

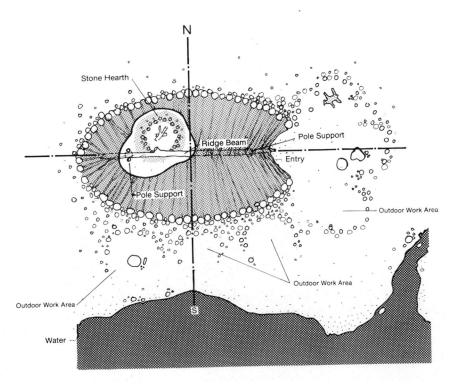

FIGURE 1-1 Paleolithic camp.

FIGURE 1-2 Stonehenge, Salisbury Plain. (Photograph by Phillip Tabb.)

is believed that Stonehenge and other similar Neolithic and Bronze Age structures were built by fairly large populations, and mysteries surrounding these efforts still exist. Refer to Figure 1-2.

Although there are numerous examples of early solar planning, several cultures have been selected for examination here. Cities that are clearly organized in relation to the sun include the city of Knossos in Crete; the city of Teotihuacán located in Mexico; the ancient Greek city of Priene located in Asia Minor; cliff dwellings of the Anasazi Indians at Mesa Verde, Colorado; and the Chinese capital city of Ch'ang-an during the Tang dynasty. Each of these cities demonstrates use of solar energy and a profound organization in response to the orientation of the sun. Both city plans and architectural designs reflect this awareness.

The Great Palace of Knossos was built and rebuilt during the Minoan civilization, which dates between 2500 and 1400 B.C. Located on the island of Crete, the palace had a long history. The actual site sloped to the south and southeast. The palace was originally built just off the cardinal points (it is oriented slightly west of south). Although it was built to a fairly high density, most of the individual units received sunlight. The main entries lead to the central court, which served the administrative and domestic quarters.

The city of Teotihuacán was located in south central Mexico. It dates between 100 B.C. and A.D. 700. At its golden period it had a population of 200,000 people. There were architectural monuments paying homage to the sun and moon. The city had a large avenue running north and south called the Avenue of the Dead, which connected the major pyramids and temples. The paths and streets were organized on a grid on the cardinal points. Here again, most of the buildings received sunlight during winter months.

The ancient Greek city of Priene is located in Asia Minor (now Turkey) and was built in the Hellenistic period around the fourth century B.C. Because of continual flooding, the original city was abandoned for a site that was located on higher ground. The 4000 residents moved to the foot of nearby Mycale, a promontory that had varied topography sloping to the south and southeast. Despite the varying topography, the city was laid out on a grid on the cardinal points. The streets were oriented in a similar way to the earlier Olynthian street plans, which put major terraced streets along the east-west axis and minor streets along the north-south axis running up the mountain. This arrangement allowed nearly all of the buildings to have good solar access. Figure 1-3 illustrates the city plan.

The Anasazi culture rose, flourished, and then vanished over a period of approximately 1300 years. The Anasazi settled in the high desert region of southwestern Colorado around the time of Christ. Eventually they migrated south and east. It is speculated that they were either

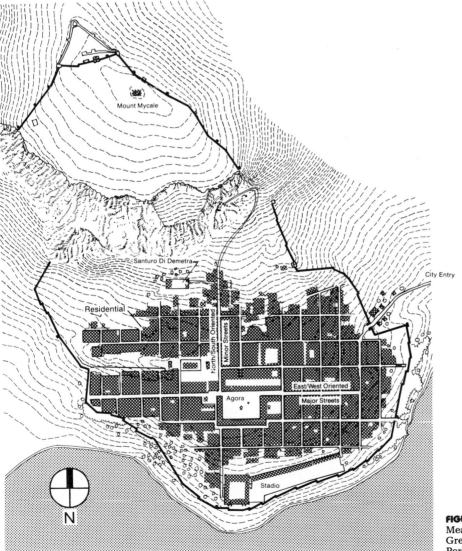

FIGURE 1-3 Pianura Del Meandro, City of Priene, Greece. (From Butti and Perlin.[8])

FIGURE 1-4 Mesa Verde, Colorado. (Photograph by Jane Harris.)

starved or driven away by someone else. At any rate, this culture left behind an extraordinary testament to their achievements in the stone structures at Mesa Verde. According to David Muench and Donald Pike, authors of *Anasazi: Ancient People of the Rock,* "The Anasazi were builders and settlers on a large and permanent scale, and it is for this that they are best remembered."*

The ruins at Mesa Verde begin to tell a story about a way of life in this semiarid environment. Life was not just confined to the cliff dwellings. It also extended throughout the plateaus and mesas where the Anasazi gathered wood for building and fire, found spring water, and grew corn, which was their staple diet. Extreme weather conditions probably necessitated withdrawal to the cave for protection. In the winter, the large cave roof provided shelter from snow, and in summer, it provided shading from the hot midday and afternoon sun.

A close look at the physical form of the community reveals a remarkable ability to mitigate extreme weather. The cave measures nearly 500 feet in the east-west direction and 130 feet deep, and it arches 200 feet high. The juxtaposition of the building structures in relation to the form of the cave suggests a clear understanding of the movement of the sun. Mesa Verde is located near 37° north latitude and, therefore, has a winter midday altitude of approximately 29.6°. The brow of the cave permits sunshine to enter the cave during the winter mornings and afternoons, and this warms the rock and buildings alike. In spring and autumn the brow begins to shade the buildings at midday, when temperatures are

*Taken from *Anasazi: Ancient People of the Rock* by David Muench and Donald G. Pike.[44] Copyright © 1974 by American West Publishing Group. Used by permission of Crown Publishers, Inc.

usually warm, and permits penetration in early morning and late afternoon, when temperatures are often cooler. In summer the brow provides complete shading where the rock and buildings remain cool despite higher outdoor temperatures. Figure 1-4 illustrates the cave dwelling found at Mesa Verde.

Many solar-oriented cities built over a vast period of time existed in China. The entire city of Ch'ang-an, which was the capital of China during the Tang dynasty in the fifth century, was built on a rectangular grid oriented on the north-south axis. The city was located inland near 35° north latitude. Marketplaces were interspersed within the grid to serve the residential areas, and the Imperial City was located near the center of the city. This plan demonstrates the importance of solar energy at the planning level.

Example after example of solar-oriented cities indicates city plans that were organized for the use of solar energy. These cities are found on most continents, and they represent developments that span thousands of years. This is true from the time of the Paleolithic camps to the more recent cities in Asia and Europe. The sun and solid wood fuel formed an energy basis that structured and patterned life for this incredibly long span of time. As we moved in time toward the Middle Ages and eventually to the industrial revolution, city form took on a new order generated by new energy sources, primarily fossil fuels, and the new freedoms associated with the development of these fuels.

Fossil Fuel Era

The fossil fuel era is projected to span a mere 600 years, beginning around the time of the industrial revolution in the mid-1800s and trailing off sometime in the mid-2400s. Natural gas and oil are expected to be exhausted earlier, in the 2000s, while coal will be with us a little longer, to the end of the era around 2450. Figure 1-5 illustrates the relatively short duration of the fossil fuel era in relation to other key historic events and periods of time. We are now considered to be near the halfway point of this era. In other words, nearly half of the world's known supply of

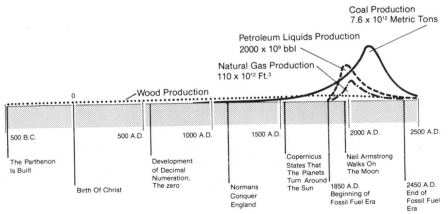

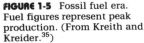

FIGURE 1-5 Fossil fuel era. Fuel figures represent peak production. (From Kreith and Kreider.[35])

fossil fuels has already been used. In slightly over 100 years of development, between 1860 and 1985, vast changes have occurred through the use of these fuels. They have affected contemporary life and the resulting city and community forms. In fact, the changes are so great that in many areas of the world there is little trace of previous patterns of development that evolved over thousands of years.

Development patterns have undergone tremendous change with the use of fossil fuels, and several characteristics are quite evident. The first characteristic is the trend toward greater urban development caused by rural families moving to the cities. The second is city forms organized around transportation and movement, especially when the automobile was introduced and made available to the average family. The third is the development of central power plants that provide electricity to every home. The fourth is more center- or radial-oriented city plans, which in many instances do not relate to the cardinal points. The last characteristic, which is of more recent origin, is the emergence of residual products of the fossil fuels themselves, such as petrochemicals and plastic products of every imaginable form. In a discussion about fossil fuels and American civilization, Wilson Clark, in his book *Energy for Survival*, states:

> America has set the pace for the lifestyle of the high-energy civilization. It was here that oil was first drilled and exploited on a large scale. It was here that coal and steam condensed a continent into a few days' train journey. It was here that the electric light and the electric distribution system were invented. It was here that abundant energy and mass production made the cheap automobile ubiquitous. It was here that the potent fire of the atomic nucleus was harnessed to produce still more energy—and destruction.*

In 1861, before the Civil War, 90 percent of the fuel consumed was wood. By 1885, coal consumption surpassed that of wood. Crude-oil production took a dramatic increase around the same time. Between 1860 and 1870, crude-oil production rose from 500,000 gallons to 2 million gallons a year. The fuel-oil market was slower in developing because fuel oil cost more and because the equipment for burning it had not yet been introduced at the larger scale. Natural gas was first used in 1824 to provide light for homes, shops, and streets in Fredonia, New York. In 1855, natural gas was used in simple burners for lighting, cooking, and some heating. Transportation underwent profound changes with large-scale networking of the railroad, especially in Europe; increased shipping; and the invention of the automobile and airplane. The introduction of the gasoline-fueled engine, which spawned the gigantic automobile industry, probably had the greatest impact upon city planning and growth. It was in 1903 that Henry Ford introduced his gasoline-powered automobile and announced the formation of the Ford Motor Company. Since then, the industry has grown into a bewildering influence upon the built environ-

*From *Energy for Survival* by Wilson Clark.[13] Copyright © 1974 by Wilson Clark. Reprinted by permission of Doubleday & Co., Inc.

ment. In fact, it has come to structure and dominate life in many industrialized cities all over the world.

A major drawback to steam, gasoline, and diesel engines is the fact that they need fuel to be transported to them for operation. Movement of great quantities of fuel involved high costs. Electric power solved the problem and could be made from the fossil fuels. With Alexander Graham Bell's patent of the telephone in 1876 and with Thomas Edison's invention of the incandescent light bulb in 1879 and central power and distribution stations in 1882, the electrical industry had taken off. When electricity reached the private home where electric outlets were provided in each room, a vast appliance business began producing washing machines, clothes dryers, dishwashers, vacuum cleaners, furnace motors, refrigerators, and an array of other machines and gadgets. The elevator and central heating and air-conditioning systems made it possible for taller and larger buildings to be built. Electricity could deliver power from steam, fossil fuels, and water to distant sites at a relatively low cost. This phenomenon gave great freedom in the organization of both new and existing communities.

CURRENT DEVELOPMENT PATTERNS

In the United States, the major settlement pattern took the form of the *metropolitan community,* which consisted of a city center and suburbs served by railroad. Industries were located in the urban areas, with the exception of the factory towns in New England and the company towns in the south. Usually surrounding these metropolitan communities were networks of agricultural market towns. The urban areas developed in similar ways.

The first development was the *center,* the heart of the region with a built-up central business district that functioned as the marketplace. The center grew more dense and extended upward. It provided many specialized services, such as banking, commercial activities, and wholesale distribution and housed supporting offices and communications and medical centers. The second development was the *frame,* which gave support to the city center. It often had a mix of commercial, warehousing, industrial, and residential uses. It was in the frame that the older inner-city neighborhoods were located. Third was *suburbia.* New political boundaries, taxation policies, the automobile, and overcrowding in the city center all added to the popularity of moving to the suburbs. And fourth was the *fringe,* which was located in between suburbia and rural areas outside of the city. Between the years 1910 and 1960 the population of the United States doubled, with a majority of people living in urban areas. A look at most twentieth-century cities and communities reveals a clear trend in development.

Modern growth of the human settlement has affected each of these four areas—center, frame, suburb, and fringe—in different ways. The push and pull between the center and the frame and between the suburbs and the fringe have caused a great deal of stress. But the most visible change in the twentieth century has occurred in suburban areas not in large city centers, despite the impact of skyscrapers. According to the *Residential Development Handbook* published by the Urban Land Institute:

> At one time the process of residential land development consisted of merely acquiring a tract, filing a plat of its division into blocks and lots, and then selling those lots to buyers. New subdivisions were usually added on to built-up areas. But a series of social, economic, and physical factors brought about significant changes: zoning ordinances and subdivision regulations were instituted as land use controls protecting public health, safety, and welfare; an economic depression brought construction to a halt; and World War II created a further housing shortage. When conditions returned to normal after the war, a building boom took place.[72]

In the United States, between the years 1946 and 1975, over 44 million privately owned housing units were constructed. The residential development was horizontal, moving outward from city centers. The key to rapid development was the subdivision process. It allowed for fairly quick turnover from raw land to developed land. Land parcels were purchased, economic feasibility was determined, the character of the development was planned, the architectural details were finalized, construction was initiated, and finally sales completed the cycle. The impact of this pattern of growth is difficult to fathom. According to Wilson Clark, "It is the development of energy resources that determines the limits of all growth."[13] Thus, development during this time was due to both the increase in population and the availability of cheap electricity, natural gas, and other liquid fuels.

FIGURE 1-6 The American subdivision process: (a) the township, (b) the agricultural section, (c) the quarter-quarter section, and (d) the city block.

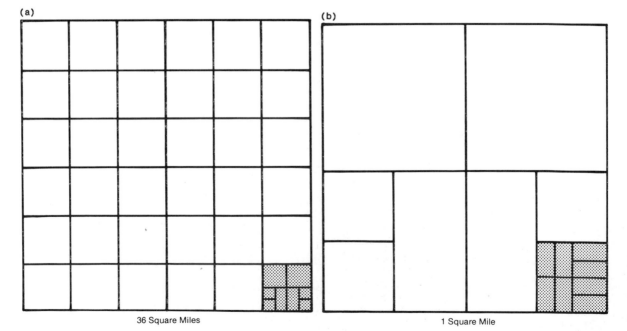

(a)

36 Square Miles

(b)

1 Square Mile

Subdivision

To settle the midwestern and western parts of the United States after the Louisiana Purchase (in 1803), a simple and expedient system of land development was designed by Thomas Jefferson. It involved a standard method of subdividing the land into parcels oriented on the cardinal points of the compass. A large grid spread across America, and it is still extremely visible today in the modularized landscape. This grid can be traced from Ohio to California. Various scales of land plots were possible. First was the *township,* with 36 square miles measuring 6 miles by 6 miles. Second was the *agricultural section,* with 640 acres measuring 1 mile by 1 mile. The subdivision process began with the agricultural section, and each successive division resulted in smaller and smaller plots of land. Therefore, we had the quarter section, with 160 acres; the half-quarter section, with 80 acres; and the quarter-quarter section, with 40 acres. This was the smallest subdivision for agriculture purposes.

For community-scale development, the quarter-quarter section was further divided. It went from 20 acres, to 10 acres, and to 5 acres, which was a unit of land that approximated the American city block, the gross dimensions of which measure 330 feet by 660 feet. The city block was further subdivided into individual land parcels. Common dimensions range from 50 to 60 feet along the street by 140 feet in depth. Space for an alley was usually provided down the center of the block, which gave access to city utilities and services, including energy. Refer to Figure 1-6, which illustrates the subdivision process.

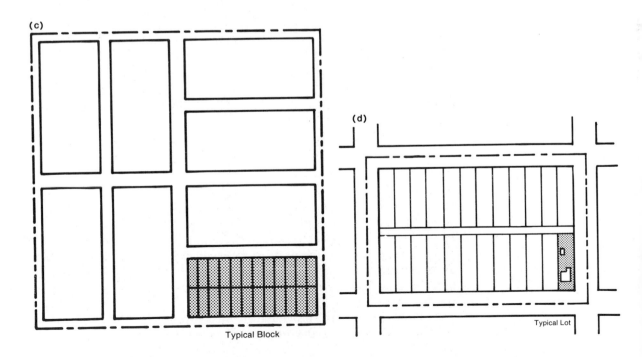

(c)

Typical Block

(d)

Typical Lot

FIGURE 1-7 Rural and suburban development. Development along the Colorado River, Grand Junction, Colorado. (Photograph by Phillip Tabb.)

Configuration of street patterns within the Jeffersonian grid obviously fell on either the east-west or north-south orientations. This street pattern was often broken when established transportation routes or natural features were dominant. River directions, for example, were more often than not aligned off north-south axis orientations. As a consequence, city-center grid patterns followed these natural orientations until the Jeffersonian grid dominated usually in the suburbs or along the fringe. Individual land parcels or lots in turn followed street orientations. The net lot areas varied but were on average between 7000 and 10,000 square feet. The last drawing in Figure 1-6 illustrates a typical lot within a block configuration.

The Jeffersonian grid provided the structure for development. It was extremely effective, but the grid, which extended endlessly in four directions, seemed boring and repetitious. So, many of the suburbs built in the fifties and sixties took on a new fabric: lacy and more free-form. It fit nicely within the mosaic of the Jeffersonian grid, yet it was quite different. It was more spacious and provided more semipublic outdoor spaces. And often it provided many amenities to the individual landowner. In the more affluent developments, homes were organized around golf courses, lakes or ponds, marinas, or recreational play areas. Integral to many of these developments were light commercial facilities and elementary schools. They began to provide a more self-sufficient lifestyle, with residential, commercial, and institutional facilities organized in close proximity to one another.

Often greenbelts surrounded these developments protecting them from external noise, automobile fumes, and hazardous traffic. They also provided an aesthetic separation. The greenbelts were also used to isolate commercial and industrial land areas from residential areas. Open spaces on the perimeter that were not suitable for housing or commercial development were often made into greenbelts. They often were a part of flood plains or other natural land features. Major freeway systems were planned adjacent to these emerging free-form developments. Figure 1-7 illustrates a development in Grand Junction, Colorado, with adjacent greenbelt and highways.

Residential development, as it is practiced today, has not changed much; however, many of the steps in the process are more articulated. Generally there are five steps: (1) market analysis and land acquisition, (2) preliminary planning and design, (3) final planning and design, (4) construction, (5) marketing and sales. Although there are many home builders and developers in the United States, most of the housing stock is constructed by a few large developers. Their work accounts for approximately 80 percent of the market. Most builders are smaller and provide somewhere around ten to twelve housing units per year. The larger developers are more involved with mass-produced homes, while the smaller ones tend toward the more customized homes.

Residential Energy-Use Patterns

After the Second World War, energy supplies in the United States were plentiful. There was a great deal of coal, oil petroleum, and natural gas. There was a small amount of electricity generated through hydroelectric plants and nuclear energy. This was the era of the "golden medallion" home, or the all-electric home. All you needed was a hookup to the centralized power grid. Nothing could be more simple. Energy was not a major planning determinant. According to Wilson Clark, "The assertion that Americans are reliant on electrical energy is an understatement."[13] Home appliances include electric resistance heating systems, electric hot-water heating systems, electric ranges, air conditioners, freezers, dishwashers, refrigerators, and portable appliances. Today, residential energy consumption is substantial, taking about one-fifth of the national total for all sources of energy and about one-third of all electricity gen-

TABLE 1-1
Residential and Commercial Energy Consumption, 1978[49]

Country	Energy Use, Million Tons of Oil Equivalent	Share of Total National Energy Use, Percent	Energy Use per Person, Tons of Oil Equivalent
United States	442.62	33	2.01
Canada	46.32	33	1.95
Sweden	12.95	38	1.56
Netherlands	21.04	39	1.50
West Germany	78.89	39	1.29
France	50.76	35	0.95
United Kingdom	45.01	31	0.81
Italy	31.80	30	0.56
Japan	56.59	21	0.49

erated. Table 1-1 presents residential and commercial energy consumption in 1978. The table relates energy use, percent of national use, and use per person in the United States and eight other industrial countries.

In the United States, space heating in temperate or cold climates and space cooling in warm climates are the greatest energy end uses for residences. Hot-water heating is second, followed by refrigeration, food preparation, freezing, and other uses. The category of "other" uses of energy is increasing at a phenomenal rate. Appliances covered in this category include lights, a myriad of kitchen appliances, home computers, televisions, radios, and stereo equipment, to name a few. The trend toward greater and greater energy use is reinforced by the high-energy lifestyle, which is characterized by a social structure that promotes a kind of synthetic individualism supported by consumption. To a certain degree, it replaces cooperation and sharing at the level of the home and neighborhood with high-energy products.

Since the oil embargo of 1973, this pattern has changed a little. Energy conservation has been taken seriously by many individuals. A great deal of human energy has gone into the study of energy. In America energy consumption has been reduced through the development of new (especially high) technology: solar collectors, air-to-air heat exchangers, movable insulation, thermal storage, thermostatic control systems, and solar thermal electrical development. Two groups seem to be involved: concerned individuals and large corporations with federal funds for research and development. It is ironic that a 4000- or 5000-square-foot single-family house can be considered energy-efficient just because it has a solar energy system. It seems to be a case of not being able to see the forest for the trees.

Existing and new homes have been made more energy-efficient through the incorporation of many conserving techniques and devices. However, this has not substantially changed the thrust of American life. This is probably due to a lagging economy, insulation from our energy problems, and the difficulty to effect change in a large way. People are still consuming vast quantities of energy even with the emerging energy consciousness. It all seems to lead to the following questions: Do we really have an energy problem? If so, do we really have the ability to change? And what will be the cost?

ENERGY-RESPONSIVE DEVELOPMENT

Despite the perplexing problem of energy consumption in the United States, many developers have stepped forward in an attempt to redefine suburban development. As early as 1941, solar housing and subdivision were demonstrated. In the suburb of Glenview, near Chicago, a solar-oriented subdivision was built by developer Howard Soan. Today, energy considerations continue to affect both the housing units as well as site-planning concepts. Energy conservation in concert with solar energy collection make up the major design scenarios. The developments are discriminating to the extent that they are much more energy-efficient, and generally the quality of construction is far superior to that of only a few years ago. A major problem facing most of these new developments centers around the economics and marketing that are involved. With increasing inflation, construction costs have risen dramatically. Many of the energy considerations are also costly and compound this situation, rendering the homes unsalable. In spite of the difficulties, many of these pioneering projects have been realized and with great success.

Several single-family detached houses and multifamily developments have been designed and constructed within the last decade. The projects, in most instances, are located in favorable locations, that is, they are located in fairly populated areas where the climate is temperate or cold (and there is therefore a need for heating) and where there is a fairly high amount of solar radiation available during the winter months. The first projects are located in Santa Fe, New Mexico; the second in Davis, California; and the third in Boulder, Colorado. This is not an inclusive listing of projects but represents the range of project types currently being developed in the United States.

The Santa Fe, New Mexico, Examples

Three projects developed by Wayne Nichols and Susan Nichols in Santa Fe represent an incremental approach to solar energy planning with a gradual increase in scale. The first project, called First Village, was developed in the early 1970s on a 40-acre parcel of land 6 miles south of Santa Fe. It was conceived as a low-density, one unit per five acres, cluster of eight single-family homes. The selling price of each unit was oriented toward the luxury market, in the $100,000 to $170,000 range. The project was considered the first speculative solar development in New Mex-

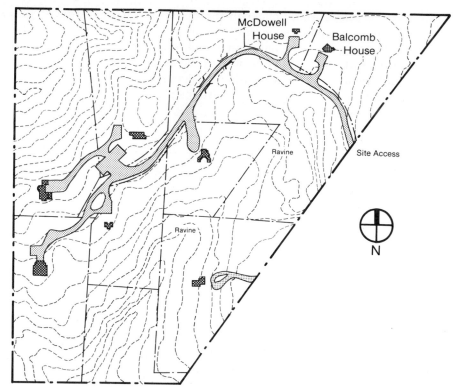

McDowell
House

Balcomb
House

Ravine

Site Access

Ravine

N

FIGURE 1-8 First Village site plan. This early solar development features eight individually designed passive solar homes. (From Lumpkin and Nichols.[39])

ico. Planned in phases, the first two homes demonstrated both passive and active solar heating techniques. The last six homes employed only conservation and passive techniques.

A photograph of the Balcomb House, designed by architect William Lumpkin along with Susan Nichols, shows the large solar greenhouse in an L-shaped plan. Inside the solar greenhouse are masonry storage walls and two heat-recovery systems that charge two rock beds located in northern zones of the house. Refer to Figure 1-9. Figure 1-10 is an isometric of the McDowell House, which has active space heating and rock storage. The housing forms vary, but the styles are similar. They represent the regional pueblo style with grafted solar technologies. Thick adobelike walls; curving forms; flat roofs with vegas; light, simple interiors; corner fireplaces; and natural landscaping are characteristic of the architecture in the southwest. Sizes of the homes of First Village vary from 1650 to 2400 square feet. Each is uniquely designed and integrated to its specific site. The houses proved marketable even though they were experimental in nature. Solar technologies could, indeed, be wed to a strong regional style of architecture.

The solar technologies used in this early development are still considered to be the state of the art for residential use. The active solar systems simply employed the water-cooled flat-plate collectors with storage for several days. Collectors generally were placed high on the flat roofs where

FIGURE 1-9 Balcomb House photographs; top, exterior and left, interior. (From Lumpkin and Nichols.[39])

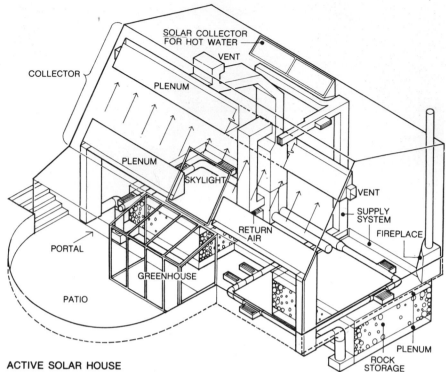

SOLAR COLLECTOR
FOR HOT WATER

COLLECTOR

VENT

PLENUM

PLENUM

SKYLIGHT

VENT

SUPPLY
SYSTEM

RETURN
AIR

FIREPLACE

PORTAL

GREENHOUSE

PATIO

PLENUM

ROCK
STORAGE

FIGURE 1-10 McDowell House isometric. Active solar house.

ACTIVE SOLAR HOUSE

they had good solar access and were out of view. Except for the first home, they were limited to domestic hot-water heating. The passive systems varied from house to house. Direct-gain systems, in the form of either vertical south-facing glass or clerestories, were used extensively. Thermal walls or Trombe walls were used occasionally. A thermal wall has a thermal mass, usually concrete or concrete block, directly behind vertical glass. The sunspace or solar greenhouse, such as the one illustrated in Figure 1-9 on the Balcomb House, was also used in this development. Generally the homes were set into the ground 3 or 4 feet to reduce height and gain additional insulation with the earth. Active components were often incorporated with the passive systems to augment thermal storage in below-grade rock beds. Despite the experimentation in design and construction of these early solar homes in Santa Fe, energy performance is exceptional. These early successes have led to greater confidence and more ambitious projects.

La Vereda was the second project in this progression of developments by the Nichols. They used an open-space community concept. In a planned unit development, nineteen homes were built on quarter-acre lots inside the city limits of Santa Fe. The homes were organized into three compounds or clusters, and each house was planned for good solar access and south orientation. Two fairly large ravines separate ridges upon which the houses were built. Refer to Figure 1-11 for the develop-

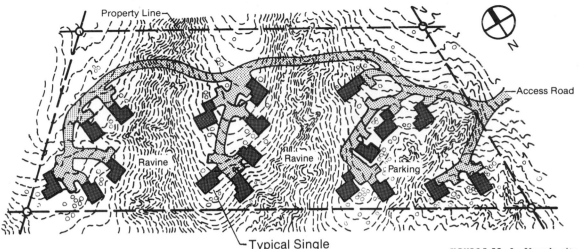

Property Line

Access Road

N

Ravine

Ravine

Parking

Typical Single
Family House

FIGURE 1-11 La Vereda site plan. Located in Santa Fe, New Mexico, these nineteen houses are on three ridges. Each house is oriented south.

ment site plan. Open space adjoins each house, and natural vegetation was preserved, as in the previous development. The development was aimed at a lower-cost market, with homes priced between $80,000 and $110,000 (1978 dollars). The units were considerably smaller than those built in First Village. The two- and three-bedroom houses varied from 1100 to 1950 square feet in size.

The care that was involved in the site planning, unit design, and construction is apparent. The development goes beyond being merely a statement of energy techniques. It is responsive to the spirit of the land, the sense of community, and the human factors that are terribly important, especially to many who live in Santa Fe. This is a good example of the mix between the concerns associated with marketing and cost and the demands associated with solar technologies and energy-conservation techniques. The blend is not obtrusive but rather strong and sympathetic to the architectural feeling characteristic of the region.

To help reduce costs, the developers attempted to standardize some of the processes. Four house designs were created along with two basic hybrid systems that were both active and passive. Nearly all of the homes were depressed into the ground 3 feet on the north, east, and west sides. The first system type was a solar greenhouse with either mass walls or rock beds for storage. The second system type was a concrete thermal wall with direct-gain windows. Both systems include active domestic hot-water heating systems. The combination of methods provides between 70 and 90 percent of the annual requirements for heat. (This, of course, is subject to the ways in which individual users actually interact with the systems and to their energy-use patterns.) This performance is quite good considering that Santa Fe experiences 5567 annual heating degree-days. Figure 1-12a illustrates a group of four solar houses within this development, and Figure 1-12b shows an individual home. Although there is

(a)

(b)

FIGURE 1-12 La Vereda housing; above, group photograph and right, individual home.

a variety of architectural forms, there is certainly a continuity of style common to the southwest.

The third Nichols development is located in a more urban setting near the center of old Santa Fe. The development is called La Vereda Compound, and it is their most recent project. On a 5-acre south-sloping site, twenty-six new condominium residences, which have been designed by architect Ed Mazria and Susan Nichols, are mixed with nine existing buildings. The density is seven units per acre. Three basic unit designs have been integrated to duplex and quadruplex building forms, and they have been planned in harmony with the existing single-family homes. Costs for the condominiums are around $250,000 (1982 dollars). Figure 1-13 shows the site plan identifying the new and existing buildings, and Figure 1-14 shows the character of the development.

Having completed three solar development projects, the Nichols have

Sloping Topography

Townhouses

Existing
Structures

Parking

Parking

Property Line

Parking

N

Townhouses

Access Road

FIGURE 1-13 The Compound site plan.

FIGURE 1-14 The Compound project photograph.

Orchard

Typical Single
Family House

Bicycle Paths

Common

East/West
Oriented Streets

Community
Gardens

Common

Community
Building

Commercial
Building

N

Community
Gardens

FIGURE 1-15 Village Homes
site plan. Adapted from
Bainbridge et al.[5]

evolved an energy-oriented planning, design, and construction process.
Key to the continued success of these projects, in their opinion, is the
strength of the home association. The condominium ownership calls for
a home association that will be responsible for the maintenance of all the

FIGURE 1-16 Village Homes aerial photograph. (From Corbett.[15])

common areas. Watering, snow removal, and general maintenance will be handled by the association. Strict solar-access protection is also provided under the jurisdiction of the home association. (Refer to Appendix B for a solar-access covenant.)

The Davis, California, Examples

Paralleling the early Santa Fe solar work was the Village Homes subdivision in Davis, California. From the initial planning stages in 1972 to the first ground breaking in 1975, the development has evolved into an extremely interesting demonstration of energy-oriented community planning and solar architecture. It was planned for the development of a sense of community as well as for the conservation of energy and natural resources. Refer to Figures 1-15 and 1-16.

The subdivision was planned on a 70-acre site that was elongated along the north-south axis. Arterial access streets run parallel to the site in the north-south direction. Feeder streets into the development are in the east-west direction. One large common was planned near the center of the development with a variety of commercial and outdoor activities. To help create a further sense of community, the planners were careful to form neighborhoods that encouraged interaction. This was accomplished with many convenient and useful common areas that are accessible by automobile, bike, and foot. Access streets are at the perimeter, and a pedestrian path runs through the center. The homes are positioned to create a common space with varied activities.

In an organized yet informal way, the 200 solar homes have been planned at an approximate density of 3 units per acre. Nearly all of the homes in the development have been constructed. Most of the homes were designed and built by Michael Corbett and John Hofacre. In contrast to the developments in Santa Fe, the homes of Village Homes are oriented to a lower-cost market. The home costs vary but are consistently in the $40,000 to $60,000 range (1979 dollars). This is considerably lower than the cost of the Santa Fe examples. Floor areas vary from 1000 to 1750 square feet; this size is slightly below average for single-family houses presently built in the United States (approximately 1700 square feet). The types of solar systems used were similar to those used by the Nichols. Direct-gain windows, sunspaces, clerestories, and distributed and concentrated thermal masses were employed. The systems were often owner-built and sometimes lacked sophistication but, nevertheless, functioned very well. Typically, the houses of Village Homes save 50 percent more energy than houses in neighboring developments.

Davis, California, experiences 2800 heating degree-days (base 65°F) and 1100 cooling degree-days. This is considerably milder than the climate of Santa Fe. As a result, the solar systems are not as dominant. Many energy-conservation techniques were used to reduce both the heating and cooling loads. Additional insulation for roofs, walls, and foundations was carefully installed, and movable insulation for windows was often used. Active hot-water systems were also used extensively throughout the development. In fact, nearly every home has one. Characteristic of many of the homes are water walls made of vertically positioned culverts filled with water. They add further heat storage capacity to the homes.

The Boulder, Colorado, Examples

Two projects in Boulder, Colorado, represent distinctly different energy-oriented development approaches. The first one, Wonderland Hills Development, was primarily a demonstration of active solar systems by one developer and designer team. However, more recently, this developer has moved into passive or hybrid designs for both single-family and multifamily housing. The second project, Greenwood Commons, was primarily a demonstration of passive systems with the involvement of a number of developer and architect teams. The climate of Boulder is similar to that of Santa Fe, with 5460 heating degree-days. It is temperate, with winter underheating and some summer overheating. Therefore, the design of the solar energy systems has a fairly important role in the planning and architectural processes.

The solar homes in the Wonderland Hills Development were a part of an ongoing development process. Originally the homes were single-family and catered to a luxury market. Many homes were custom-designed

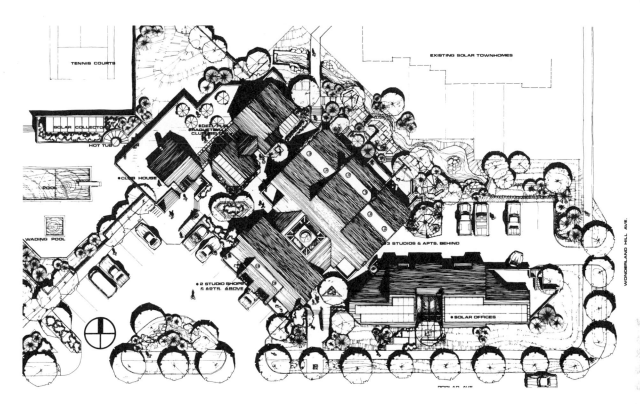

Labels within figure:
TENNIS COURTS
EXISTING SOLAR TOWNHOMES
SOLAR COLLECTOR
HOT TUB
SOLAR FRAGILE TIA? CLUB SPACE?
CLUB HOUSE
POOL
WADING POOL
3 STUDIOS & APTS. BEHIND
2 STUDIO SHOPS 5 APTS. ABOVE
SOLAR OFFICES
WONDERLAND HILL AVE.

FIGURE 1-17 Wonderland Hills site plan. The Wonderland Hills solar development was part of cycle 2 of the HUD solar demonstration program. Courtesy of Downing Leach & Associates, Architecture-Planning-Engineering, Boulder, Colorado.

and -built. Subsequent phases moved to higher densities with a mix of single-family and multifamily residences. After the oil embargo in the early 1970s, the developers, Downing and Leach, launched their operations into solar energy with the houses that were then under construction. As a part of the second cycle of the U.S. Department of Housing and Urban Development demonstration program, the first solar homes utilized active solar collection systems for space and hot-water heating. Figure 1-17 identifies the location of the solar-heated community center, with shops and offices, in the center of the development. Figure 1-18 is a photograph of the original active solar demonstration. The latest efforts feature passive solar single-family residences on zero lot lines as pictured in Figure 1-19. This house was a part of the Solar Energy Research Institute (SERI) and National Association of Home Builders (NAHB) solar demonstration. The two-story house has a full basement, detached garage, and active domestic hot-water heating system.

The residences of Greenwood Commons are all two-story single-family detached houses. Most of the houses have been developed by individual designer and builder teams. The overall master plan was planned by architect Ervin Bell. The homes are in the $170,000 range (1981 dollars). Thirty-eight individual lots are served by a curving north-south street. In the center of the development is a home-owners' common. Many of the homes were a part of a project jointly sponsored by SERI and NAHB. What

FIGURE 1-18 Development photograph.

FIGURE 1-19 Prototype passive solar house.

FIGURE 1-20 Development photograph.

is unique about the project is the demonstration of the variety of energy-conservation and solar energy techniques. Among the systems types that were demonstrated are direct-gain systems, thermal walls, water walls, double envelopes, superinsulation, sunspaces, and a number of energy-conservation measures. Typically, the systems deliver 60 to 70 percent of the annual heating demand. Although most of the sites have been developed to date, many of the homes have still not been sold.

The systems are integrated in a variety of ways, and the level of design and the quality of construction are generally high. Although the development is successful from a solar-systems point of view, the homes lack continuity of style and sensitivity to one another—each home seems to compete with the others. As a result, the homes tend to force an overwhelming introversion where privacy, instead of community, actually dominates. There is little homogeneity. In isolation many of the house designs are quite exciting. Figure 1-20 is a photograph taken from the southwestern corner of the development. There are many architectural forms and styles. In fact, there is little in common. Some houses slope to the north; some slope to the south; and some have vertical solar walls. The house at the far end is made with stucco, others with wood; and in Figure 1-22, there is a large dominant metal roof. Figure 1-21 illustrates the Greenwood Commons site plan with a rather conventional suburban layout.

SUMMARY

History has an interesting way of repeating itself. By looking at this brief historical sketch of residential development and solar planning, it is apparent that solar energy was once an important planning determinant. Despite the present-day problems of inflation and the high costs for energy and construction, the use of solar energy is growing in popularity. It is not merely a fad but something more fundamental than that. It is a kind of clearing or centering of architectural design and planning. It is giving direction and meaning in a changing time—perhaps a postmodern

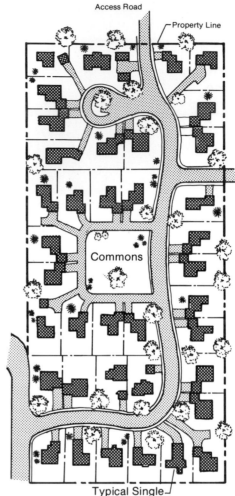

Access Road

Property Line

Commons

N

Typical Single Family House

FIGURE 1-21 Greenwood
Commons site plan. Many
houses within the Greenwood
Commons development were
a part of the SERI and NAHB
demonstration program.
(Courtesy of Ervin J. Bell,
Boulder, Colorado.)

FIGURE 1-22 Typical solar
house.

time. What is healthy with this generation of solar building development is the commitment to make it work. Solar planning determinants are actually being considered and tested within the conventional planning process.

In 1978 there were 2,022,000 housing starts in the United States for a construction volume of $110.2 billion. This represents a great deal of housing. For solar energy to dramatically impact this building volume, residential settlement makers will need to consider the limitations of solar energy in the creative development process. Current housing demands and the need to reduce energy consumption can both be met. Residential development can respond to the energy needs and limitations of solar energy characteristic to each region.

LIMITATIONS 2
OF
SOLAR ENERGY

In nature there are fixed limits for summer and winter, day and night, and these limits give the year its meaning.   THE I CHING*

Solar energy use in buildings is increasing at a phenomenal rate. There are many reasons for the popularity of solar space heating, domestic hot-water heating, and daylighting. Solar energy is free. In many areas of the United States solar technology is extremely cost-effective. Because solar energy is dispersed, it is available to a great number of people located all over the world. It is nonpolluting, and it is a renewable resource. The fact that thousands of solar projects are realized each year is evidence that home owners, home builders, and developers are changing to energy-oriented design. New and old buildings alike are being adapted to a variety of energy-conserving and solar energy techniques.

This trend is not without its drawbacks. Initially, "going solar" can be costly, time-consuming, and frustrating. Many of the solar products are new and, therefore, have not been proved over time. Very often lifestyle adjustments are required in order to best realize energy savings. As larger residential settlements are planned and as more complex buildings are designed, the need for detailed limits is more critical. Pressure from tremendous growth in the industry is beginning to affect the development process in positive ways. To ease the transition to greater solar energy

*From Wilhelm.[75] The Richard Wilhelm translation of *The I Ching or Book of Changes,* rendered into English by Cary F. Baynes. Bollingen Series 19. Copyright 1950, © 1967 by Princeton University Press. Copyright © renewed 1977 by Princeton University Press. Excerpt, p. 231.

use, the limitations of solar energy need definition and clarification. To quote Farrington Danniels, author of *Direct Use of the Sun's Energy:*

> The direct use of the sun's radiation is not new, but we have new materials to work with, such as inexpensive transparent sun-resistant plastics and semiconductors of high purity. We have new ideas in science, accumulated experience in engineering, and a broader knowledge of worldwide opportunities and needs. Thus it is important to reexamine all the ways in which science and technology can help to make direct use of the sun practical.[17]

The purpose of this chapter is to focus on the limitations of solar energy so they can be considered at the planning level. By understanding the source of these limitations—whether they are related to natural laws, technology, cost, or design—more intelligent decisions may be made. Four areas have been chosen for investigation: *limitations of climate, limitations of solar technology, limitations of cost,* and *limitations of scale.*

LIMITATIONS OF CLIMATE

Solar energy is customarily defined in terms of electromagnetic wavelengths and is measured in langleys per minute or Btu's per hour. At the outermost surface of the earth's atmosphere, 2.0 langleys per minute have been measured. This is the *solar constant.* Recent discoveries have shown the solar constant to vary slightly, but for the purpose of planning and design, this has a negligible effect. What happens to the solar energy as it passes through the atmosphere, however, is of importance because of its effect on climate. Dramatic changes in climate occur in seasonal, daily, and hourly cycles.

Two phenomena are major causes of weather. The first is the axial tilt of the earth. The earth's axis is tilted 23.5° from perpendicular in the direction of the sun's ray. This shift remains constant throughout the elliptical orbit around the sun and is responsible for the changes in season, or *seasonal cycles.* The second is the rotation of the earth about its own axis over an approximate 24-hour period. This rotation, of course, causes the daily warming and cooling, or *diurnal cycles.* The physical characteristics of climate form an important beginning for the design process. The low intensity of solar radiation at the surface of the earth, the intermittent quantity of energy received over time, the changes in the ambient environment, and the necessity for conventional mechanical systems during periods of extreme weather form a basis for climate-responsive design.

Solar Energy

The first limitation of solar energy is its low intensity or low flux density. By the time solar energy reaches the earth's surface, it is relatively weak. As a result, large collection areas are necessary for large-scale utilization. The larger the collection area, the more expensive and the more difficult it becomes to integrate the systems into buildings. Solar energy reaches the surface of the earth by direct radiation and diffused radiation. The amounts of radiation vary geographically. As mapped by M. I. Budyko, distribution of solar radiation tends to occur in higher levels around the equator and in lower levels around the poles.[7a] There is more energy available around the equator between latitudes 25° north and 25° south. Toward the poles the intensity falls off. The direct radiation is dependent upon the position of the sun relative to the surface of the earth and the clearness of the atmosphere. Diffused radiation is caused primarily by the scattering of dust particles in the atmosphere and water droplets in clouds. Even greater dispersion of solar radiation occurs in urban areas where there are high levels of air pollution.

Second of the limitations of solar energy is its intermittency. The intermittent behavior of solar energy happens in several ways. During regular daily cycles, half of the earth is in darkness, and during the regular annual cycles, the seasons change. A large percentage of the world's population exists between latitudes where climate is temperate. In these climate regions there are usually high percentages of cloud cover. As much as half of the earth at any given moment is covered by clouds, and clouds are most often found throughout the temperate zones. The combination of these effects can cause extreme weather changes.

Figures 2-1 and 2-2 show the mean daily solar radiation throughout the United States for the months of June and December. The difference in these values accounts for one of the ways in which solar energy is considered intermittent. In Salt Lake City, Utah, for example, the daily total solar energy for the month of June is 621 langleys, while in December it is as low as 146 langleys. The difference between the summer and winter quantities of solar radiation is 475 langleys, which is a factor of over 4 times as much energy during the summer month. It is quite evident that the uneven distribution of solar energy and its intermittency are important in the design of solar systems and the determination of conservation measures.

Weather

Extreme weather conditions are difficult to predict in terms of both frequency and intensity. Therefore, design decisions have been generalized for these extremes and buildings depend upon conventional heating, ventilating, and air-conditioning systems to provide comfort during these periods. Once nonsolar systems have been sized for a particular weather condition, such as extended periods of hot or cold weather or extended periods of cloudiness or precipitation, they simply require additional dosages of fuel, usually natural gas or electricity. Solar systems do not work this way. In extremes, a solar system cannot be turned on for "additional dosages." At best a solar system can provide energy for only several days, largely because of the economics involved. There is a diminishing return on the investment of a solar system. The collection and storage components of the system are too costly to be sized for the extremes. Under the

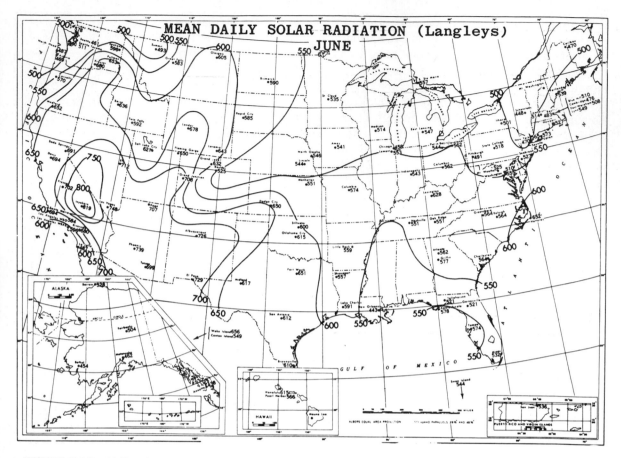

MEAN DAILY SOLAR RADIATION (Langleys)
JUNE

FIGURE 2-1 Mean daily solar radiation, June. (From U.S. Department of Commerce.[65])

present conventional fuel cost structure, the additional cost for these solar components is prohibitive. Therefore, solar systems are sized for between 50 and 80 percent of the total load depending upon the specific climate and economic criteria for which they are designed. This means a continued reliance upon conventional fuels for peak-load energy demands.

Weather can differ from cities to rural areas. The urban environment, with its built forms and materials, can cause changes in the weather and microclimate in many different ways. Solar radiation, temperature, wind speed, humidity, cloudiness, and precipitation are all affected by the urban environment. Table 2-1 lists these climatic characteristics and delineates the average differences between urban and rural areas. Two figures that stand out are the winter reduction of solar radiation and the annual reduction of wind in the urban area. Another figure worth mentioning is the increase in winter fog also associated with the urban environment.

For the purposes of planning and design, the climate of a particular region can be divided into two categories: *prevailing climatic, or macro-*

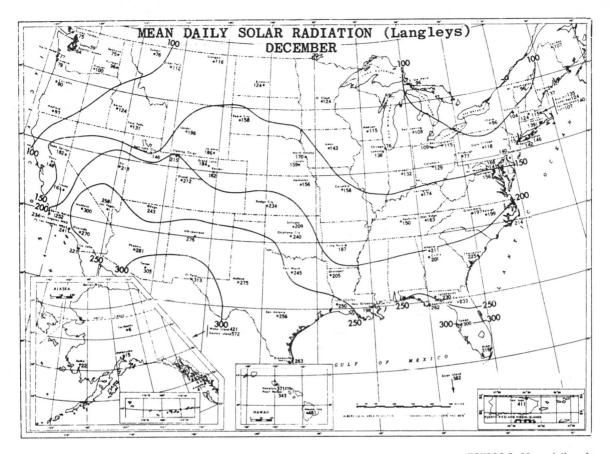

FIGURE 2-2 Mean daily solar radiation, December. (From U.S. Department of Commerce.[65])

TABLE 2-1
Climatic Changes Due to Urbanization[36]

Element	Parameter	Urban Compared with Rural
Radiation	On horizontal surface	−15%
	Ultraviolet	−30% (winter), −5% (summer)
Temperature	Annual mean	+0.7°C
	Winter maximum	+1.5°C
	Length of freeze-free season	+2 to 3 weeks (possible)
Wind speed	Annal mean	−20 to −30%
	Extreme gusts	−10 to −20%
	Frequency of calms	+5 to −20%
Humidity (relative)	Annual mean	−6%
	Seasonal mean	−2% (winter), −8% (summer)
Cloudiness	Cloud frequency and amount	+5 to 10%
	Fogs	+100% (winter), +30% (summer)
Precipitation	Amounts	+5 to 10%
	Days (with less than 0.2 in)	+10%
	Snow days	−14%

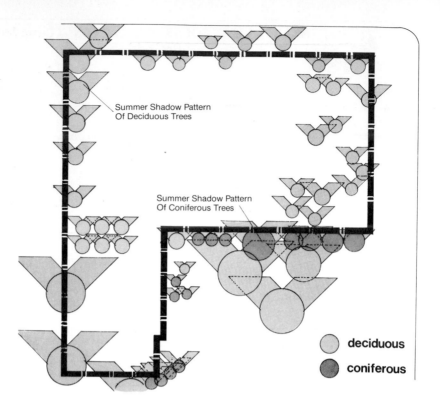

Summer Shadow Pattern
Of Deciduous Trees

Summer Shadow Pattern
Of Coniferous Trees

deciduous

coniferous

FIGURE 2-3 Summer site analysis. (Student Project, University of Colorado at Boulder, Jeff Davis.)

climatic, conditions and *microclimatic conditions.* The prevailing conditions, obviously, refer to the overall climatic characteristics of the region, while the microclimatic conditions refer to the local characteristics. They are identical or at least similar where local conditions do not alter the prevailing ones, but sometimes they differ drastically. A microclimate can exist that is quite unlike the prevailing average conditions, or macroclimate. It is important to understand the distinction here because larger development projects may have building sites that demand very different microclimatic design approaches. The analysis that follows in Figures 2-3 and 2-4 illustrates these two climatic categories for both summer and winter by simply looking at the effects of sun shading and wind upon a site. Note that shadows of deciduous trees in winter have been reduced. This reduction depends upon the branch density of the specific trees. However, the branch and needle density of the coniferous trees is high and is cause for concern. The shadows of both the deciduous and coniferous trees have been identified for both times of the year.

Many factors may influence the weather patterns of a microclimate. Most of them are difficult to quantify without on-site scientific equipment with which data can be taken over a reasonable period of time. This can be an expensive and time-consuming process. For the most part, micro-

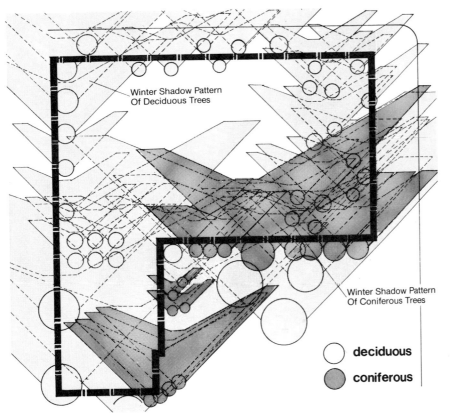

Winter Shadow Pattern
Of Deciduous Trees

Winter Shadow Pattern
Of Coniferous Trees

○ deciduous

● coniferous

FIGURE 2-4 Winter site analysis. (Student Project, University of Colorado at Boulder, Jeff Davis.)

climatic design is accomplished through analyzing the prevailing climatic data, where it exists, and intuiting the microclimate effects, usually sun and wind.

Comfort

Understanding the limits of comfort is extremely important in climate-responsive design. The relationship between comfort and both prevailing and extreme weather needs to be examined for each project. Thermal comfort varies from activity to activity and from culture to culture. For example, sleeping is vastly different from skiing. The limits of comfort with the activity of sleeping are more restrictive than with the activity of skiing. With the latter activity, temperatures can be lower, air movement can be greater, etc. Between cultures, temperature settings for various activities will vary. Table 2-2 illustrates temperatures for various activities in three different countries: the United Kingdom, Germany, and the United States.

Comfort levels are not fixed. In the United States, for example, energy conservation has increased over the last decade, albeit slowly. Conservation seems to be linked to the availability of energy. As energy becomes more scarce, conservation increases. The annual growth rate for the use

TABLE 2-2
Comparative Recommended Temperatures (°F) in U.K., Germany, and U.S.*

	U.K.[†]	Germany	U.S.[‡]
Living room	65°	68°	
Bedroom	60°	68°	
Bed-sitting room	65°		73–75°[§]
Kitchen	60°	68°	
Bathroom	60°	71.6°	
School			
Assembly halls	65°		68–72°
Gymnasium	55°	59°	55–65°
Teaching rooms	65°	68°	72–74°
Changing rooms		71.6°	65–68°
Hotels			
Bedrooms	60°		75°[¶]
Ballroom	65°		65–68°
Dining room	65°		72°
Theater auditorium	65°		68–72°
Hospital	65°		72–74°
Operating room	65–70°		70–95°
Shops, stores	65°		65–68°
Offices			
Conference rooms		64.4°	
Typing rooms		68°	
Circulation		59°	

*From Rapoport.[50] © 1972 by Basic Books, Inc., Publishers. Reprinted by permission of the publisher.
†Institute of Heating and Ventilating Engineers (IHVE) guide 1965.
‡American Society of Heating, Refrigerating, and Air-Conditioning Engineers (ASHRAE) guide 1963.
§U.S. homes. Note (1) no differentiation between rooms and (2) major difference with U.K., especially bedrooms.
¶Major difference.

of energy in the United States dropped over 3 percent from the early 1970s to the late 1970s.

The climatic effects upon human physiology can be graphically presented, and the limits of comfort can be identified. The bioclimatic chart plots dry-bulb temperature, relative humidity, and air movement, giving a range or zone in which we experience "comfort." At higher temperature ranges and at higher humidity levels, the velocity of winds is indicated in feet per minute. Figure 2-5 is a bioclimatic evaluation for a moderate climate zone in the United States (Grand Junction, Colorado). Monthly weather data have been superimposed upon the chart illustrating times of the year in which weather conditions fall out of the range of comfort. This kind of analysis forms a basis for the thermal design procedures that

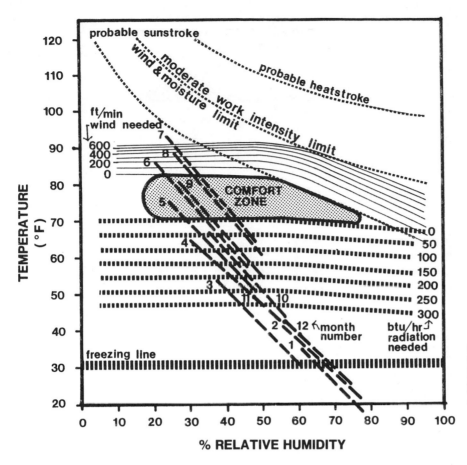

FIGURE 2-5 Bioclimatic evaluation. Walker Field terminal building, Grand Junction, Colorado. (Courtesy of J. F. Kreider and Associates.)

may be used for a project in this location. Note that most of the year is out of the comfort zone, thus requiring a great deal of supplemental heating and cooling. The chart has its limitations. This is not a true energy picture because it does not indicate microclimatic conditions, it does not consider cultural or behavioral characteristics, and it does not identify potential internal gains. It does give an indication of the months of the year that require space conditioning.

For the most accurate analysis of the effects of weather, hourly data should be used. This allows for both seasonal and diurnal changes in solar radiation, cloudiness, air movement, humidity, precipitation, etc. The limitations of climate can then be cross-checked with program, site, and other user requirements for the appropriate architectural and planning responses.

LIMITATIONS OF SOLAR TECHNOLOGY

Popular definitions of solar technology fall into two major categories: *passive systems* and *active systems.* Common to both are fundamental com-

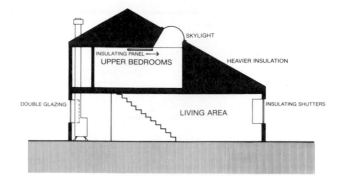

CROSS SECTION

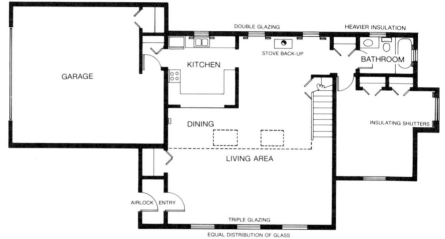

FIGURE 2-6 Energy-conservation approach.

GROUND FLOOR PLAN

ponents of collection, storage, distribution, and control. Active systems usually have a more tightly knit package combining all of the components into a comprehensive manufactured system. Passive systems, generally regarded as being more simple, have the same components, but they are more integrated into the form of the architecture. The limitations associated with the collector are the same for both system types, but the relationships (that is to say, the heat-transfer methods) among components vary, especially with the passive systems, where they generally play a dual role with the architecture. Active systems are usually more isolated and function in a more controlled way. Passive systems are less mechanical (sometimes defined as nonmechanical) and often require human interaction in order to function most efficiently.

Energy conservation is a prerequisite to the design and engineering of passive and active systems. Well-targeted conservation measures can reduce energy consumption tremendously and can limit stress upon the solar systems. Conservation affords a great deal of flexibility in design.

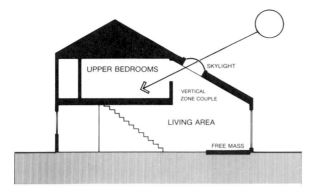

CROSS SECTION

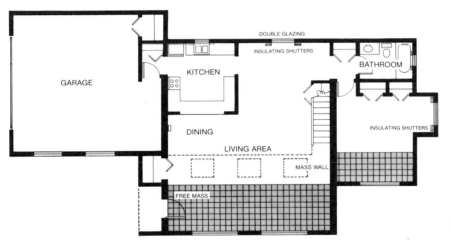

GROUND FLOOR PLAN

FIGURE 2-7 Partial solar-collection approach.

There are many techniques, and there are many ways in which to conceal their impact upon the architecture. For example, Figure 2-6 illustrates an energy-conservation approach in a floor plan and section designed for a temperate climate region. Figure 2-7 illustrates a partial solar collection approach (solar-tempered) to the same design, and Figure 2-8 illustrates a full passive solar collection approach. The degree to which solar energy is considered in the design of a project can be expressed in incremental levels. A continuum exists between conventional design and full solar energy design. The limitations of technology, therefore, exist within this continuum, and they grow with the complexity and importance of each of the conservation and solar energy measures. In *The Passive Solar Energy Book*, Edward Mazria writes:

> Architecture in the twentieth century has been characterized by an emphasis on technology to the exclusion of other values. In the built environment this concern manifests itself in the materials we build with, such as plastics and synthetics. There is an existing dependence on

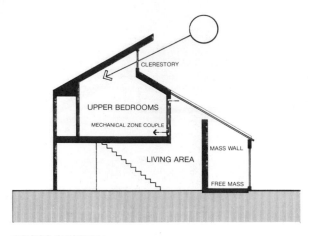

CROSS SECTION

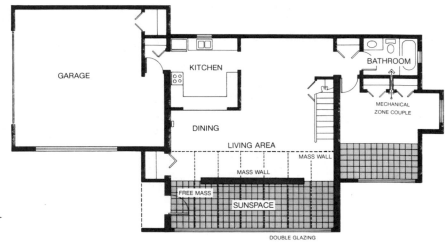

FIGURE 2-8 Full solar-collection approach.

GROUND FLOOR PLAN

mechanical control of the indoor environment rather than exploitation of climatic and other natural processes to satisfy our comfort requirements.*

The following outline gives energy-conservation measures for new buildings. The measures are organized into three categories: reducing inefficiencies, reducing energy demand, and reducing building heat load. Many of the measures can be applied to existing buildings. Because this book focuses on residential buildings, most of the measures deal with "skin"-dominated conservation measures.

*Reprinted from *The Passive Solar Energy Book* © 1979 by Edward Mazria.[40] Permission granted by Rodale Press, Inc., Emmaus, Pa. 18049.

1. Reducing inefficiencies
   a. Increase heating-plant efficiency.
      (1) Replace pilot lights with electronic ignition devices.
   b. Recover heat from stack or flue losses.
   c. Employ more efficient heat-distribution system.
      (1) Clean filters.
      (2) Check pumps, fans, and other motors.
      (3) Where applicable, insulate heat-distribution lines.
      (4) Clean and check effectiveness of diffusers.
   d. Incorporate the most efficient household equipment and appliances.
      (1) Compare kilowatt-hour consumption of various appliances before purchase.
2. Reducing energy demand
   a. Reduce time of use of heating or cooling plant.
   b. Employ thermal zones (when certain spaces are not in use, they can have lower temperatures in winter and higher temperatures in summer).
      (1) Group daytime activities together and isolate from nighttime activities if possible.
   c. Employ lower temperature settings in winter and higher settings in summer.
      (1) Winter temperature settings can range between 62 and 68°F— the age, health, and activity of the user will influence the temperature setting.
      (2) Tie conventional heating and cooling systems to zones.
   d. Employ nighttime temperature setbacks.
      (1) Typically, the night temperature is set back between 5 and 10°F while the house is occupied; the temperature can be lower if not occupied (check temperature needs of indoor plants, if applicable).
3. Reducing building heat load
   a. Reduce peak load demand—especially if the peak demand is used as a basis for conventional energy pricing.
      (1) Avoid excessive energy use during extreme cold periods.
      (2) Reduce use of shower, bath, dishwasher, washing machine, etc.
      (3) If using a solar domestic hot-water system, use it during recharging periods.
   b. Reduce building envelope winter heat (conduction and convection) losses and summer heat (conduction and convection) gains.
      (1) Increase building insulation levels (over previous levels) with below-grade, perimeter, wall, and roof insulation.
      (2) Use multiple glazings and/or movable insulation if applicable.
   c. Reduce building envelope winter heat (infiltration) losses and summer heat (infiltration) gains.

(1) Carefully caulk and weather-strip around all windows and doors.

(2) Insulate sole plate, behind electric outlets, around flues, ducts, and other perimeter wall-mounted fixtures and equipment.

(3) Use continuous vapor barriers.

(4) Incorporate energy-responsive landscaping to reduce the effects of direct winds in winter.

(5) Insulate basement walls, if applicable.

d. Turn off lights and appliances when not in use.

(1) If possible, employ manual override switches for various systems of lighting for convenience.

(2) Place in convenient locations (perhaps on a zone-by-zone basis).

(3) Place at appropriate height for children.

Incremental Integration

Energy design techniques are difficult to prescribe at the national level even though there are some universal techniques that can be generalized regardless of climatic zone. Energy design is dependent upon an extremely complex set of factors: climate, market, program, cost, level of user participation, reliability of local energy industry, prevailing architectural styles, and aesthetics, to name a few. In some instances energy-conservation measures have been able to affect broad markets throughout the United States. As demonstrated by many innovative developers and builders, full solar energy projects have been realized with great fervor. Many projects have used state-of-the-art passive and active systems. However, not all projects have used the full solar approach. Many developers have chosen to ease into energy-oriented design. As the scale of the development increases, there is greater hesitancy to use unproven systems, and the economics often dictate less energy-efficient schemes. The trend is to make energy-efficient changes at a rate that is appropriate to the needs of each project. The concept of incremental change is an appealing one, especially at the local level, because it has the potential of reaching the broadest spectrum of users.

For the purpose of this chapter, the integration increments are divided into six major parts. They are (1) conventional designs, (2) energy-conservation designs, (3) sun-tempering designs, (4) incremental passive designs, (5) full passive designs, and (6) active designs. The progression, from conventional to active systems, in most cases is additive. In other words, the more energy-oriented the approach, the more the accumulation of energy techniques. For example, full passive designs will often include increments of mass, sun tempering, and energy-conservation measures. Figure 2-9 illustrates these increments and shows their relationship in the progression. Descriptions of the increments follow.

1. Conventional Designs. There are no visible energy features. Residential projects usually have an equal distribution of windows or glass, no additional mass beyond standard interior partitions and materials, and no orientation constraints.

2. **Energy-Conservation Designs.** Conservation choices are nonsolar, including higher mechanical efficiencies for conventional heating and cooling systems, reduced energy demand, and reduced building load.
3. **Sun-Tempering Designs.** These include energy conservation, window relocation with increased south aperture and orientation constraints, reduced north glazing, and no additional thermal mass.
4. **Incremental Passive Designs.** These include energy conservation, increased south aperture, increased thermal mass, correct interior zoning, and incremental cooling techniques.
5. **Full Passive Designs.** These include energy conservation, large south aperture, reduced north glazing, optimally placed thermal mass, overnight thermal storage, correct interior zoning, and passive cooling.
6. **Active Designs.** These include energy conservation, heat recovery, domestic hot water heating, space heating, and photovoltaics.

Conventional Systems   Conventional single-family houses currently built in the United States are fairly predictable. The average size is between 1600 and 1700 square feet. The house form varies with two-story, one-story, bi-level, and split-level designs. Half of the homes have full or partial basements, and a majority of them have either one- or two-car garages. In the average house approximately 65 percent of the energy consumed is used for space heating, 17 percent for domestic hot-water heating, 9 percent for food preparation, and 9 percent for the running of other electrical appliances, including air conditioners. The conventional house generally has an equal distribution of windows or glass around its perimeter. That is, each side of the house has a similar total area of windows or glass. This affords a great deal of siting flexibility. The aspect ratio, or plan shape, is rectangular and not exaggerated. This includes L-shaped plans as well. A majority of homes are fueled by natural gas; most of the rest are fueled by oil or electricity. Houses in remote areas use either propane or liquid gas. In warmer climates a majority of the conventional houses have either air-conditioning or evaporative cooling. A large drawback to the conventional house is its lack of energy conservation and solar heating.

Energy-Conservation Systems   The first step in making the conventional house more energy-efficient is through the use of energy-conservation measures. Energy conservation, here, is considered to be nonsolar. The three major conservation categories are reducing inefficiencies, reducing energy demand, and reducing building heat load. The first category focuses on the heating plant of the building: employing more efficient furnaces and boilers, reducing stack or flue losses, and making the distribution system more efficient. The second category refers to reducing

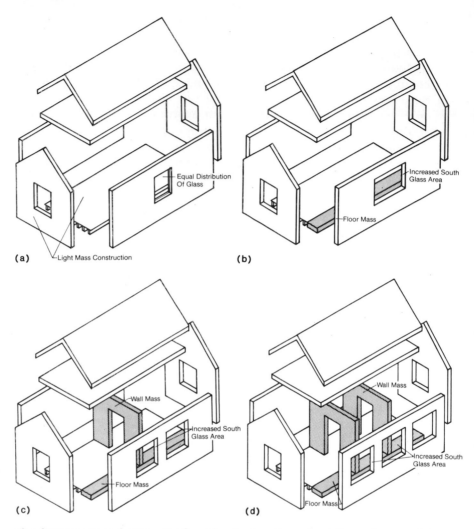

(a) ——Equal Distribution Of Glass

——Light Mass Construction

(b) ——Increased South Glass Area

——Floor Mass

(c) ——Wall Mass

——Increased South Glass Area

——Floor Mass

(d) ——Wall Mass

——Increased South Glass Area

——Floor Mass

the frequency and amount of use by the heating plant through organizing activities that require extensive energy use, employing thermostat night setbacks (between 5 and 15°F) and using lower temperature settings (say between 62 and 68°F). The third category focuses on reducing the building load by increased foundation, wall, and roof insulation; infiltration control; tighter weather-stripping, caulking, and sealing; and employing higher construction standards. The energy-conserving house has great design flexibilities and can be adapted anywhere. In situations where solar energy is not available, a superinsulation approach can be employed in which all of these techniques are used and exaggerated.

Sun-Tempered Systems Sun tempering is the most fundamental solar-heating approach. It is generally the least expensive solar design approach. It involves using energy-conservation measures and relocating windows and glazings so that areas on the north are reduced and areas on

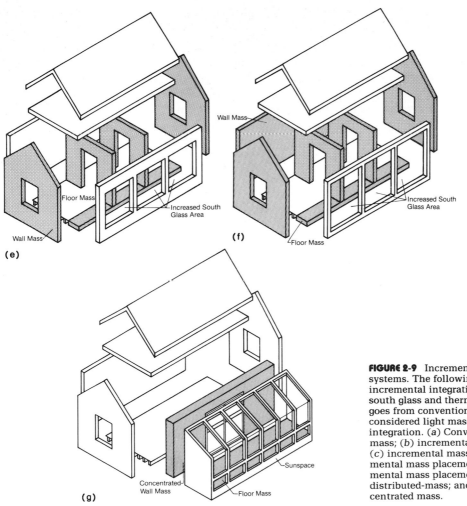

(e)

(f)

Wall Mass

Floor Mass

Increased South Glass Area

Wall Mass

Increased South Glass Area

Floor Mass

(g)

Concentrated Wall Mass

Floor Mass

Sunspace

FIGURE 2-9 Incremental integration of passive systems. The following illustrations show the incremental integration of both increased south glass and thermal mass. The progression goes from conventional construction, which is considered light mass, to full passive solar integration. (*a*) Conventional design with light mass; (*b*) incremental mass placement 1; (*c*) incremental mass placement 2; (*d*) incremental mass placement 3; (*e*) incremental mass placement 4; (*f*) full direct-gain, distributed-mass; and (*g*) full sunspace concentrated mass.

the south are increased (for the northern hemisphere). The increased south glazing area is usually accomplished with conventional methods: vertical fixed and operable windows and sliding glass doors. Spaces located behind the south glass are heated directly by the solar energy. There is no additional thermal mass in a sun-tempered house other than that which is provided by conventional stud walls and floors. Therefore, there is little night storage. In colder climates movable insulation is recommended to reduce excessive heat loss at night. This approach is relatively flexible but must conform to the solar orientation constraints for adequate solar gain.

Incremental Passive Systems With incremental passive design there is increased south glazing associated with increased thermal mass. This approach is generally referred to as a direct-gain distributed-mass system. Spaces are directly heated by solar energy, and night storage is accomplished with the thermal mass distributed throughout the architectural form (typically in walls and floors). As more solar heating is desired, larger areas of south glass and volumes of thermal mass are required. The thermal mass is sized for night storage and daytime overheating. Like sun-tempered buildings, incremental passive buildings are phototropic and need southern orientation. The more passive the building becomes, the more critical the orientation constraint. This approach is also preceded by appropriate energy-conservation measures.

Full Passive Systems   Full passive solar designs are usually distinguished by more dominant glazing and storage systems. There are three generic types: the full direct-gain distributed-mass system, the thermal-wall system, and the sunspace system. There are many variations in the design of these system types. The thermal-wall system includes Trombe walls and water walls where the mass is placed adjacent to the glazing surface. The sunspace or attached greenhouse actually encloses space. The glazing is either vertical or sloped, and the storage is often in the floor, in the north wall, or in rock storage beds. Some passive systems incorporate mechanical distribution systems although they are not strictly passive. Full passive systems characteristically have large collector areas in temperate and cold climate regions.

Active Systems   Active solar energy systems have mechanical and electrical components that help improve their thermal efficiency. They can deliver fairly high temperatures depending upon collector design. Typically, flat-plate solar collectors can produce temperatures in excess of 200°F. Concentrating collectors can produce even higher temperatures in bright sun. Storage is accomplished with water storage tanks, rock beds, or phase-change materials. To ensure higher performance, active solar collectors should be positioned at the optimal tilt angle, which varies depending upon latitude and actual energy need. At the residential scale, hot-water heating, because of its year-round demand, is the most cost-effective use for an active system. Photovoltaics or photoelectric cells are just now being commercialized for residential use; however, demonstration is still limited to a few houses.

   For the solar approaches, collection is the most limiting function of the system. Collection is area-intensive and expensive. Project sites are often oriented in a direction other than south, causing difficulty for proper collector orientation. Optimal active collector array tilts are often greater than roof angles, causing possible architectural problems. Typically collector tilt angles range between 40 and 60° from horizontal. Where collector costs are high, as in the case of photovoltaics, these limitations need careful consideration.

   Figures 2-10 through 2-13 illustrate a passive solar house designed for a temperate climate with approximately 7500 degree-days. (Refer to Figure A-9 in Appendix A for a computer-aided-design illustrating the sun's-eye view of the house on December 21.) The active collector area is 96 square feet, the sunspace area is 378 square feet, and the vertical south-

FIGURE 2-10 Bramwell House, Florissant, Colorado. Courtesy of Phillip Tabb Architects.

facing glass area is 264 square feet. This represents a total collector area of 738 square feet—a fairly sizable area for the 3150-square-foot house. Therefore, extreme care should be taken for the array orientation, tilt, and integration into the architecture. All major spaces in the Bramwell House have east-west zoning. In the center of the house are the kitchen, dining, and hot-tub areas and the greenhouse—all receiving sunlight from the sunspace and bounded by north-south mass walls. The active solar collectors are integrated into the sunspace form. A heat-recovery system pulls warm air from the top of the sunspace to two rock beds. One rock bed is positioned beneath the bed in the master bedroom and the other is beneath the living room. Both rock beds are actively charged and passively discharged (radiant release).

The sloped glazing over the sunspace uses triple-glazed heat mirrors. The total resistance value for the glazing is 4.3. This is quite high considering standard double-glazing has a resistance value of 1.8 to 2.0. While solar radiation transmission (short-wave radiation) is reduced, the resistance to heat loss has an overall benefit to the sunspace system. The active collectors are manufactured by Novan in Boulder, Colorado. The north-south mass walls are 8-inch-thick concrete block walls that are solidly grouted.

In incremental passive designs and in sunspace designs, there is a necessary relationship between solar glazing and thermal mass. In the thermal wall, of course, the relationship is a direct one; solar energy penetrating the glass will strike the thermal mass. However, the other passive

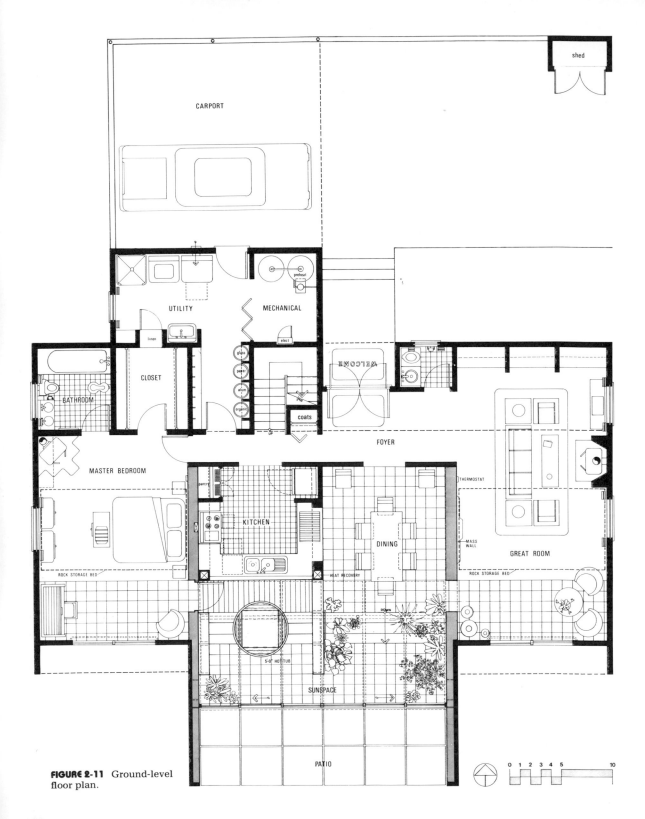

shed

CARPORT

UTILITY

MECHANICAL   preheat

elect

WELCOME

BATHROOM   CLOSET   glass   coats   FOYER

linen   paper

alum

organic   UP

MASTER BEDROOM   pantry   thermostat

KITCHEN

DINING   MASS WALL

GREAT ROOM

ROCK STORAGE BED   HEAT RECOVERY   ROCK STORAGE BED

DOWN

5-0" HOT TUB

SUNSPACE

FIGURE 2-11 Ground-level floor plan.

PATIO

0 1 2 3 4 5   10

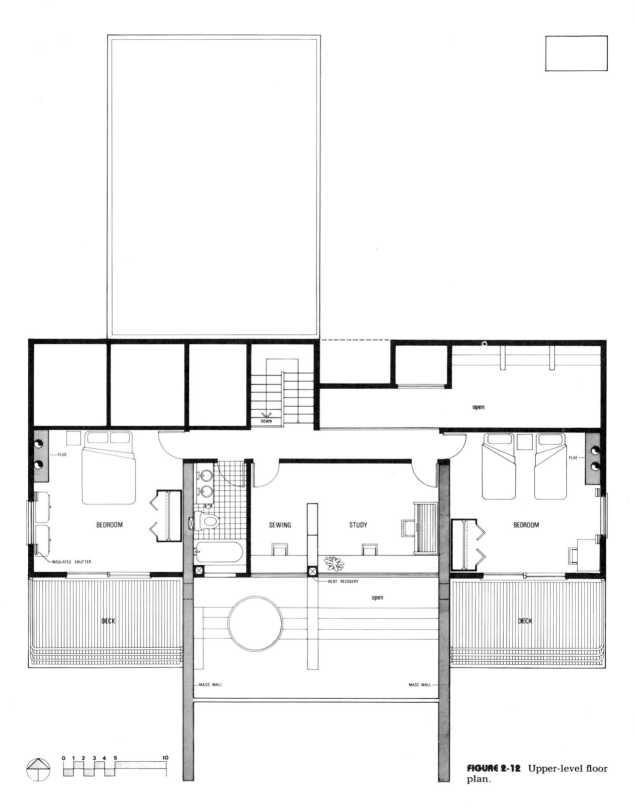

BEDROOM

FLUE

INSULATED SHUTTER

DECK

open

DOWN

SEWING

STUDY

HEAT RECOVERY

open

MASS WALL

MASS WALL

BEDROOM

FLUE

DECK

0 1 2 3 4 5 10

FIGURE 2-12 Upper-level floor plan.

51

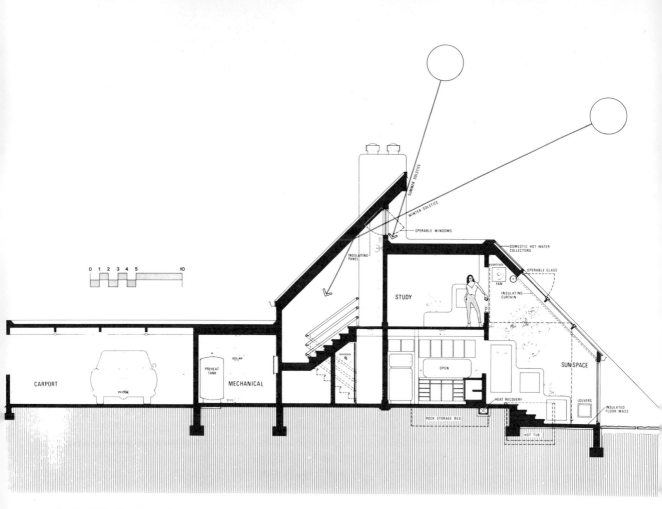

FIGURE 2-13 North-south section.

designs often have difficulties in exposing mass, which is often deep within the building, to direct solar radiation.

The project shown in Figure 2-14 is the new airport terminal building for Grand Junction, Colorado. This project is a cycle 3 U.S. Department of Energy demonstration of commercial application of passive solar technology. Figure 2-15*a* and *b* illustrates several computer-aided drawings of the project showing the relationship of clerestory arrays to the building and the relationship of the heat-recovery system and the floor mass to the collector. As can be seen, protecting solar access to collectors and to thermal mass is important at both the planning and architectural levels. In complex projects this is a more difficult task.

LIMITATIONS OF COST

One of the most limiting factors in the conversion of solar energy into use for buildings is the cost associated with the conservation and solar technologies. It is not the energy source that is costly; solar energy is free. It

FIGURE 2-14 Walker Field terminal building, Grand Junction, Colorado.

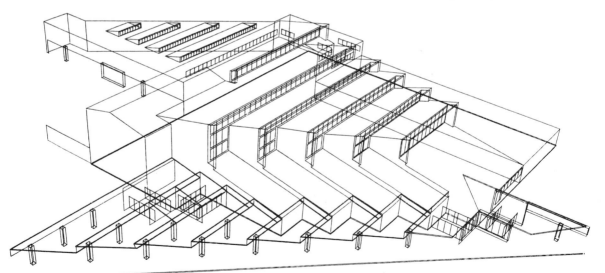

FIGURE 2-15 (a) Walker Field terminal building. Computer drawing.

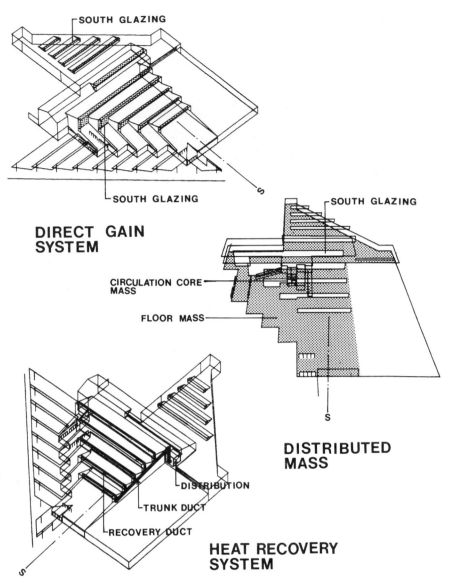

SOUTH GLAZING

SOUTH GLAZING

DIRECT GAIN SYSTEM

SOUTH GLAZING

CIRCULATION CORE MASS

FLOOR MASS

S

DISTRIBUTED MASS

DISTRIBUTION

TRUNK DUCT

RECOVERY DUCT

S

HEAT RECOVERY SYSTEM

FIGURE 2-15 (*b*) Walker Field terminal building. Drawings showing relationship of glazing to mass. [Figures courtesy of J. F. Kreider and Associates, Phillip Tabb Architects, and Nelson Greene of the Computer-Aided Design Laboratory (CADLAB).]

is the solar energy systems that incur the costs. The annualized cost of solar heating must be less than conventional heating for it to be economically viable.

The issue of cost for solar energy is centered around the additional capital that is often needed to realize the energy technologies that are not usually accounted for in construction budgets. The logic behind life-cycle costing is attractive, but it still does not contribute to the capital needed initially. Even with state and federal tax credits, initial costs are often not relieved. Both the tax credits and the energy savings are accrued over a given period of time. It can take as long as 5 to 20 years to realize the returns on energy savings.

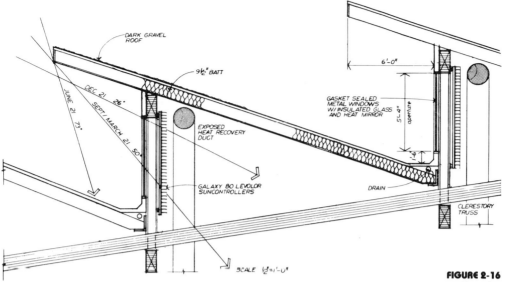

Initial Cost

The argument that solar systems, especially the passive ones, are inexpensive is true for certain measures. With new buildings, such measures as the orientation of the building, the design of overall shape, and the location of windows to the south probably do not add any cost. However, it would be naive to think that additional insulation, airlock entries, active solar systems, etc., do not add cost. They do add cost and often as much as 10 to 15 percent of a construction budget. In many cases it is difficult to isolate the energy measures from the rest of the building. Figure 2-16 is the overhang designed for the Walker Field terminal building in Grand Junction, Colorado. An estimate of major component costs, excluding federal or state sales tax, is in Table 2-3. The total cost for the overhang is $10,485. The overhang of approximately 600 linear feet is for 3000 square feet of glazing area. This amounts to $17 per linear foot.

TABLE 2-3
Overhang Cost Estimate*

Glue-laminated beams (extended)	$ 4,060
Pine roof deck	2,541
Trim at edge	748
Built-up roofing	1,620
Contingency (10%)	899
Contractor's overhead and profit (7.5%)	617
Total cost	$10,485

*Based upon 3000 ft² of clerestory glazing and Summer 1980 costs.

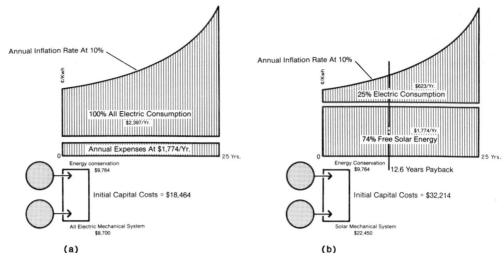

100% All Electric Consumption
$2,397/Yr.

Annual Inflation Rate At 10%

¢/Kwh

0

Annual Expenses At $1,774/Yr.

25 Yrs.

Energy conservation
$9,764

Initial Capital Costs = $18,464

All Electric Mechanical System
$8,700

(a)

Annual Inflation Rate At 10%

¢/Kwh

$623/Yr.
25% Electric Consumption

$1,774/Yr.
74% Free Solar Energy

0

25 Yrs.

Energy Conservation
$9,764

12.6 Years Payback

Initial Capital Costs = $32,214

Solar Mechanical System
$22,450

(b)

FIGURE 2-17 Life-cycle cost analyses (over a 25-year period) for a student apartment building, cycle 1, HUD demonstration program, 1975. (a) Conventional system; (b) solar-assisted system.

It is clear from the overhang cost analysis for Walker Field that the initial costs can be hidden within other building costs. In this case additional roof structure, roofing, flashing, soffit material, and finishing were necessary for a simple overhang. For projects with tight construction budgets, energy measures with the best cost-benefit ratio must be looked at first.

Not all projects have an ideal situation for the use of solar energy. Many sites slope north, have severe shading problems caused by neighboring buildings, or are located in extreme climates. Buildings often have unfavorable orientations or complex internal needs. These and other factors can lead to more expensive solar solutions. The limitations of initial cost can have a tremendous impact upon what is feasible for a given project. According to Wayne Nichols, "We know that the passive solar home costs more and anybody who says it doesn't has never built one."*

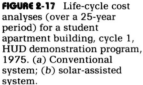

Life-Cycle Cost

Aside from the problems of initial cost, solar energy measures do have cost benefits. They accrue in time as the initial cost is paid off and conventional fuel costs rise. The lower the initial cost and the higher the fuel inflation, the greater are the energy savings. The length of the payback period is a function of the initial cost, interest on the energy systems, conventional fuel cost, and inflation. The payback period on a domestic

*Wayne Nichols quoted in W. Lumpkin and S. Nichols.[39] Reprinted from the April 1979 issue of *Progressive Architecture*, copyright 1979, Reinhold Publishing.

TABLE 2-4
Cost Comparison of an All-Electric System and a Solar-Assisted System[*,306]

1. Capital investment	
a. All-electric system	$18,464
b. Solar-assisted system	32,214
2. Cost of capital	
a. $18,464 × 0.1102 × 25 =	50,868
b. $32,214 × 0.1102 × 25 =	88,750
3. Operating cost	
a. $2397 × $\dfrac{(1 + 0.075)^{25} - 1}{0.075}$ = $2397 × 67.98 =	162,948
b. $623 × 67.98 =	42,352
4. Total costs (item 2 + item 3)	
a. $50,868 + $162,948 =	$213,816
b. $88,750 + $42,352 =	$131,102

[*]This economic analysis compares the costs of a 100 percent electrical installation and a 74 percent solar, 26 percent electrical installation. The terms of this analysis are a 10 percent interest rate on the initial capital investment over a 25-year period, and a fuel inflation rate for the same period of 7.5 percent per year. The resultant figures show the cost-effectiveness of the solar installation.

hot-water heating system will be around 5 years; for passive or active space-heating systems, it will be as long as 15 years. One thing to remember here is that few building elements actually have a payback at all. Most elements depreciate and eventually need replacement. Figure 2-17 illustrates two 25-year life-cycle cost analyses, one for a conventional space-heating system and one for a solar heating system.

The graph in Figure 2-17a illustrates a life-cycle cost for a conventional space-heating system for a student housing project in Boulder, Colorado. The conventional mechanical *systems* cost is relatively low. The conventional *fuel* cost, especially with fuel inflation, is quite high. As shown in the graph in Figure 2-17b, the life-cycle cost for the solar-assisted system is lower. The solar-assisted system shows higher initial costs and greater interest but considerable reduction in fuel and inflation costs. The payback period for the solar components of the mechanical system is a little over 12 years. The total life-cycle cost for the conventional system is $213,816; the total for the solar-assisted system is $131,102. Over the 25-year period, there is a total savings of $82,714 with the solar-assisted system. See Table 2-4 for a full cost comparison.

LIMITATIONS OF SCALE

The built environment is obviously created at different levels. Regional, city, community, and neighborhood planning are at the larger scale. Urban design, planned unit developments, and architecture are somewhere in the middle. And renovation, energy retrofit, interior design, and product design are at the smaller scale. Each scale seems to coexist with a set of conditions that limit its influence. Some conditions consistently fall within each of these levels, while other conditions are unique to each

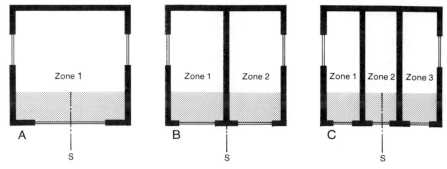

FIGURE 2-18 East-west zoning effects.

level or scale of design. Issues related to solar energy need to be identified at each scale, and extreme care needs to be taken at their interfaces.

A solar limitation common to each level is orientation. Limitations that differ on each level include development process time, use or function, number of units, construction type, density, and cost. These variables can be drastically different. For example, there can be one regional plan while there may be hundreds of thousands of a given product (for example, solar collectors). A community project can take nearly a decade to realize while an interior design project may take only several months. In order to make such a broad spectrum of limitations more manageable, three scales have been chosen for discussion. They are thermal zones, single buildings, and multiple buildings.

Thermal Zones

A *thermal zone* is a space in which comfort is affected in a particular way. In other words, a thermal zone will have a characteristic temperature, humidity, and air movement. Disregarding the input of conventional backup heating systems for the moment, most architectural designs inherently have multiple thermal zones. The more complex a design, the more thermal zones there tend to be. In fact, in a given building, there can be zones that overheat because of excess solar gain, zones that over-heat because of internal gains, zones that underheat because of no gains, and zones that underheat because of excessive heat loss. Thermal zones are generally determined by the following: relative solar heat gain, rela-tive internal heat gain, relative heat loss, exposure to prevailing winds or breezes, relative amount of daylight, and internal heat stratification or movement.

Figure 2-18 is a ground-level floor plan with simple thermal zones. The figure illustrates several adjacent thermal zones in the east-west direc-tion. Note that the plans—A, B, and C—have interior zones with south-ern exposure. Each space within each of the plans forms one zone as long as the solar energy effectively reaches all of the spaces. Plan B has two interior zones, 1 and 2, one with a western face and one with an eastern face. Plan C has three interior zones. The middle zone, zone 2, is well protected from energy loss and will experience nearly 15 percent less heat loss. Winter winds from the west could further increase heat loss for

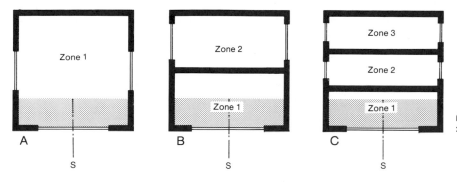

FIGURE 2-19 North-south zoning effects.

the thermal zone on the west, zone 1. Glazings located on the east and west facades should be accompanied with insulating devices. In relation to the natural forces of weather, each of these zones behaves differently. Thermal zones elongated along the north-south axis, as in plan B, will tend to be warmer on the south end, perhaps overheating, and cooler on the north end even as they are one zone. If east or west windows are desired, overheating can be lessened by placing the windows farther to the north. Note that in Figure 2-18 the window placement is in zones 2 and 3.

Orientation of zones is important for solar energy gain. In areas of high solar radiation, every zone within a plan that has good southern exposure and adequate glazing or collector area can achieve a high solar fraction. Zones placed adjacent to one another along the east-west axis will achieve this. Zones placed adjacent to one another along the north-south axis, on the other hand, may achieve different levels. Certain zones will be stressed. The southernmost zone will receive a great deal of solar energy. The southern zone, in fact, may *overheat*. The zones north of it will receive less solar energy and may *under*heat because they are only receiving east and west sun. This situation may not exist if north-south zones receive solar energy through the roof. Off-axis zone orientation may also cause a reduction in daily solar energy gain. Refer to Figure 2-19.

A thermal zone can include more than one room as long as each room is thermally coupled to an energy source. In order to achieve desired comfort levels while keeping the design fairly simple, *zone coupling* can be employed. A particular design may initially have numerous thermal zones, but by coupling many of the spaces into fewer zones, a less complex conventional system will be required. Zone coupling can be accomplished in several ways. The first way is through open planning or open architectural design. The more open the space, the greater the ability for the space to be homogeneous with a free transfer of heat and air. The degree of openness should be related to privacy needs. The second method is through mechanical means. Fans can move air from one zone to another. Although there are two distinct spaces, they act as one thermal zone. This method is particularly useful when privacy is an issue. The third method is vertical zone coupling. Spaces located above one another can be coupled as heat rises. As long as there is a thermal loop,

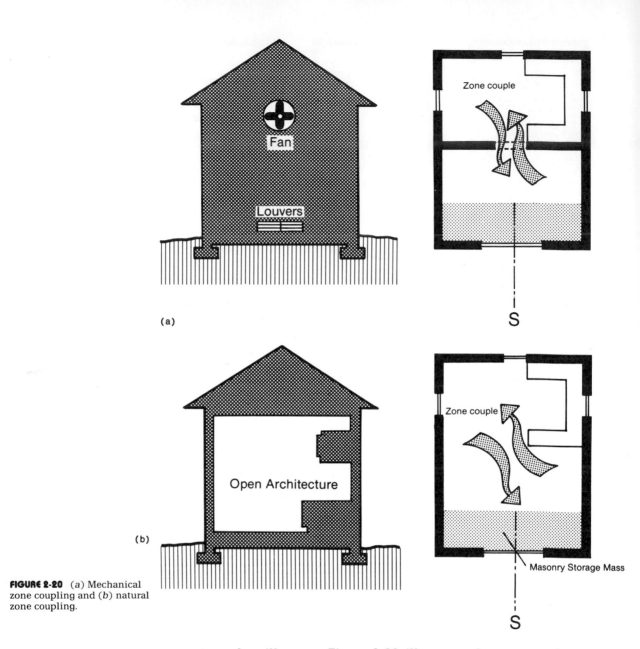

(a)

Fan

Louvers

Zone couple

S

(b)

Open Architecture

Zone couple

Masonry Storage Mass

S

FIGURE 2-20 (a) Mechanical zone coupling and (b) natural zone coupling.

a zone couple will occur. Figure 2-20 illustrates these types of zone coupling.

The results of designing for proper interior thermal zoning can be seen in Figures 2-21 and 2-22. The plan and section are for a single-family house located in the mountains. Daytime spaces, including the kitchen, bathroom, utility room, stairs, and living and dining areas, are positioned along the southern facade. In fact, some of the spaces actually project out from the major vertical southern facade. The living and dining areas, because they are also night spaces, are tucked beneath a large heavily

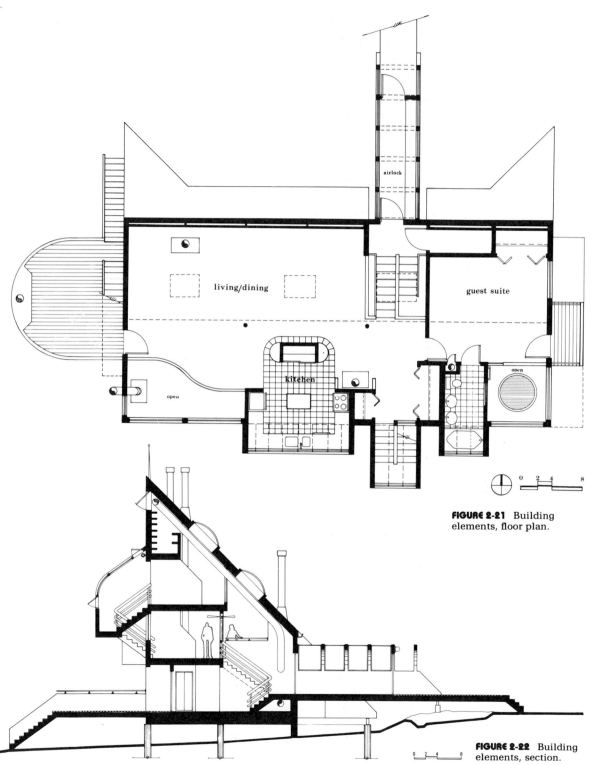

airlock

living/dining

guest suite

kitchen

open

open

0 2 4 8

FIGURE 2-21 Building elements, floor plan.

0 2 4 8

FIGURE 2-22 Building elements, section.

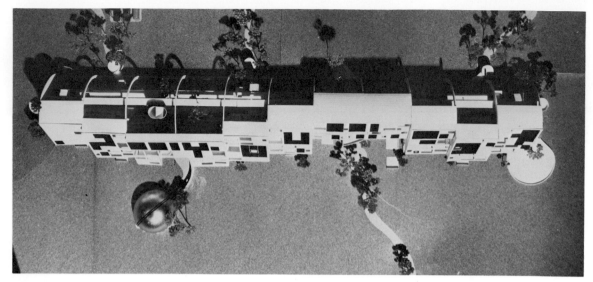

FIGURE 2-23 Building elements, model. (Photograph courtesy of Phillip Tabb, Alan Brown, and Roland Hower.)

insulated north roof. Because this house is located high in the mountains, it is considered in a cool climate with over 10,000 annual heating degree-days. In the section, three levels can be seen. They are open to one another, and with the connecting stair, they form a vertical air-circulation path, or vertical zone couple. Figure 2-23 is a model for a multifamily housing project with units placed next to one another along the east-west axis, illustrating east-west zoning.

Single Buildings

At the scale of single buildings, energy issues can be examined from two perspectives: the major and minor building forms. Major form determinants represent powerful forces. They relate to basic constraints, such as building location and orientation, relationship to the land, overall building shape, methods for solar collection, major circulation, zoning of activities, structural systems, location of entries, primary building materials, and major landscaping elements. Minor form determinants relate to more subtle constraints, including color, fenestration, thermal storage concepts, overhangs and other methods of sun control, zone coupling, interior furnishings, finishing materials, insulation values, construction details, minor landscaping elements, decoration, and "signage," or the styling of signs. The major and minor forms often overlap. There is no distinct dividing line between them. However, for the purpose of this discussion, the major form determinants are the larger-scale considerations while the minor ones are the more detailed considerations.

The major form characteristics of a building can be either dominated or barely affected by energy-related factors. The degree to which a building expresses its energy features is not necessarily a demonstration of its energy efficiency. Buildings that have the characteristic solar section, or

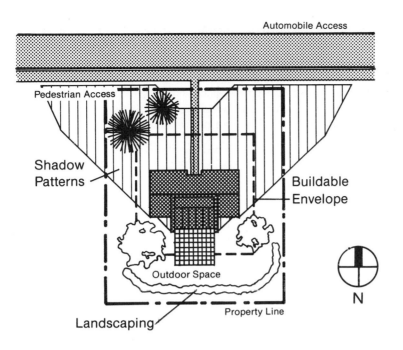

Automobile Access

Pedestrian Access

Shadow
Patterns

Buildable
Envelope

Outdoor Space

N

Property Line

Landscaping

FIGURE 2-24 Single building.

solar profile, often exaggerate the solar features. Large areas of glass to the south, often sloped, and earth berms to the north are typical of the modern solar building. A common problem with this approach is the boring north side. An extremely energy-efficient building, on the other hand, may be of conventional design and may not appear to have any obvious energy features. More residential treatment of the solar glazing details instead of a more commercial glazing system can obscure large areas of solar collection. This is becoming more true with the use of vertical collector surfaces that can be blended into conventional architectural forms. Refer to Figure 2-24 for a diagram describing considerations for single buildings. Regardless of the outward appearance of the building design, several important energy design principles should be addressed.

1. Location of building on the site
2. Relationship of building to the ground
3. Envelope design (size, shape, roof, complexity, etc.)
4. Orientation in relation to the movement of the sun
5. Aspect in relation to thermal zoning
6. Solar collection for space heating (if needed)
7. Solar collection for hot-water heating (if needed)
8. Collector protection from shading
9. Protection for major entries (from wind, glare, etc.)
10. Location of shade trees, wind breaks, and water
11. Earth sheltering (if appropriate)
12. South-lot design

The minor form characteristics are often less obvious. Yet they are just as important as the major form characteristics. The location of a thermistor for a movable insulation panel, for example, is hardly noticeable, but it could be extremely crucial to the functioning of a passive heating system. The exterior color of a building can affect its energy performance. Light-colored buildings can reduce a cooling load by as much as 15 percent. Darker colors are more favorable for buildings with a high heating demand. The following is a listing of minor form energy considerations.

1. Exterior color (value from dark to light)
2. Window location
3. Sun control
4. Energy-conservation measures
5. Thermal-mass placement
6. Zone coupling
7. Thermostatic control systems
8. Heat-recovery systems
9. Resource recycling
10. Quality of construction

Multiple Buildings

The scale of zones is limited by the thermal behavior of spaces; the scale of single buildings is limited by major and minor form modeling; the scale of multiple buildings moves into the planning realm and deals with the relationships between buildings. From an energy point of view, three issues are important at this scale: (1) solar-access protection, (2) the nature of the exterior spaces created by the buildings, and (3) the impact on the conventional planning considerations, such as zoning, utilities, street orientation, and parking. The limitations at this scale go beyond technological concerns to include social, political, and larger economic issues. The physical ramifications are not always obvious. Subsequent chapters in this book will touch on some of these areas in greater detail. Typical of the energy issues at this level are:

1. Appropriate density (based on cost, zoning, site, etc.)
2. Solar-access protection
3. Open-space design (activities, circulation, etc.)
4. Shared gardens or allotments
5. Shared energy systems (if applicable)
6. Community and privacy (indoor and outdoor)
7. Automobile access
8. Parking (automobiles, bicycles, etc. and special areas for the handicapped)
9. Street orientation and form
10. Utility distribution (recycling)
11. Architectural compatibility (style, scale, and quality)
12. Multiple building relationships (space in between)

As the scope and scale of a design or plan increase, the kinds of issues and their complexity certainly change. The design limitations at larger

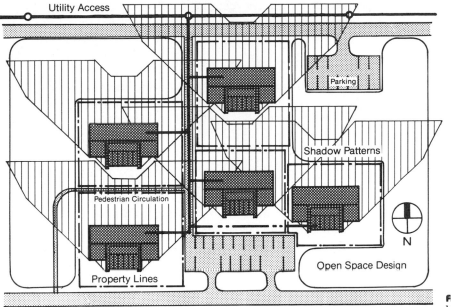

FIGURE 2-25 Multiple buildings.

Utility Access

Parking

Shadow Patterns

Pedestrian Circulation

N

Property Lines

Open Space Design

Automobile Access

FIGURE 2-26 Example photograph.

scales of city and regional planning are often only remotely related to those at smaller scales. It is important to understand the changing needs and relationships of one scale to another. Particularly, the decisions made at one scale should be sensitive to the impact upon another scale. Refer to Figures 2-25 and 2-26, which illustrate some of the limitations found at the scale of multiple buildings. They include automobile access, parking, pedestrian walkways, common- or open-space design, utility rights-of-way, and multiple-building shadow patterns.

SUMMARY

The emphasis placed upon the limitations of solar energy throughout this chapter is not intended to encourage a skeptical view of solar development. Limitations should not be construed as being negative. To the contrary, the limitations of solar energy should be seen as positive, practical boundaries within which the creative processes of energy-oriented design and planning can be nurtured. Advances in science and technology are giving us tools to better understand the nature of climate and its effects on comfort. The advances in the design professions are providing energy-efficient schemes and are enabling us to better cope with the complex problems associated with the various scales of development.

In many areas of the United States the use of solar energy is extremely cost-effective. As its use becomes more economically feasible in more areas, greater numbers of projects are going to be realized. The need to plan for solar access is extremely important to consider. Projects planned for larger populations most likely will be developed. This is going to facilitate the need for understanding and applying the central energy issues related to design and construction of solar buildings. To integrate solar energy considerations into the prevailing building process, clearer parameters are needed. The entire shelter industry needs to acknowledge the limitations of solar energy as well as the opportunities that can bring new meaning to our future communities. The task is enormous, yet we possess the resources and tools to undertake the necessary changes.

PLANNING 3
FOR
SOLAR ACCESS

*I was shown great light descending from the
heavens and great light ascending from the
earth, I watched the two come together, blend,
and become one.* EILEEN CADDY[9]

Solar-access protection is becoming more important with the increased use of solar energy in our buildings. In the past, solar demonstrations were limited, for the most part, to single-building projects, often on remote sites that had easy access to the sun. Today, many developments with multiple buildings utilize a variety of solar energy systems and require greater scrutiny in building placement and form. In some neighborhoods nearly all of the houses have been installed with solar domestic hot-water heating systems. More and more solar energy systems are being used in urban areas, where solar access is much more difficult to achieve. As a beginning step, several communities have adopted zoning ordinances that help secure solar-access protection and encourage energy-responsive design.

The concept of solar access is applicable to both existing and future buildings. New developments afford the opportunity for complete solar-access protection. It is possible to provide solar access to projects with densities as high as fifty or more units per acre. Depending upon the design, even higher densities are achievable. Providing a significant amount of solar access to the existing building stock is more difficult. According to Ralph Knowles, author of *Sun Rhythm Form*, "Our cities are nondirectional; our buildings are undifferentiated by orientation to the sun. They stand static, unresponsive to the rhythms of their surroundings."[32]

It is the purpose of this chapter to examine the issues affecting solar access. These include the *movement of the sun;* the resulting *shadow patterns* cased by building elements, single buildings, and multiple buildings; the types of *solar access;* and the relationship between solar access and *density.* Designing for solar access is fundamental to the solar energy planning process and deserves careful consideration.

SOLAR MOTION

People instinctively respond to the sun in the sky. We learn from an early age about its warmth, its light, and its intensity. We are aware of its general movement; how it is low in the morning and afternoon and how it is high at midday. We are not, however, sensitive to all of its movements, which occur minute by minute, hour by hour, and season by season. The sun's position is perceived as being relative to that of the earth's and is constantly changing. Perception of this change is now becoming more important with the advent of solar energy utilization. Without access to the sun, we cannot benefit from it. That is, we cannot use it as an energy replacement for the current nonrenewable energy sources.

We need to understand the movement of the sun for several reasons. First, we must be able to relate solar collection to solar movement. Solar collectors are usually fixed and, therefore, should be designed to maximize their collection potential. It is obvious that they need to be oriented so that they can collect the solar energy over the greatest period of time during the day and year. Second, we must understand solar movement to be able to protect from too much solar energy. In warm climates solar energy is intense, and buildings need to be designed for cooling. Even in temperate and cool climates, overheating can occur during the summer months. In fact, a solar building can overheat at midday during the winter months. Third, solar movement is important for aesthetic reasons. The quality of light changes over time, and specific effects can be achieved through careful planning and design.

To plan for the movement of the sun, certain geometrical relationships need to be known. Although the sun's movement is a dynamic process, several angles characterize its effect on a stationary object on the earth's surface. They are the declination of the sun, the latitude angle, the hour angle, the solar altitude angle, the azimuth angle, the profile angle, and the incidence angle. Refer to Figure 3-1, which illustrates most each of these angles. These angles and the interrelationships between them are fundamental to the movement of the sun and form a basis for design for solar access.

The Geometry of the Sun

The earth follows an elliptical orbit around the sun. It is farthest away on June 21, approximately 95.9 million miles, and closest on December 21, approximately 89.8 million miles. Because of the axial tilt of the earth, greater amounts of solar energy are reflected in northern latitudes. Dur-

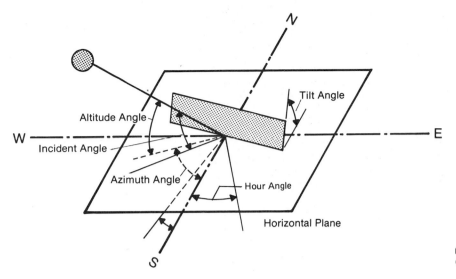

FIGURE 3-1 Solar angles.
(From Kreith and Kreider.[35])

ing winter months, these northern regions receive less solar energy even though the earth is closer to the sun at that time. During the summer months, more intense sunshine is experienced because of the longer daylight hours and the more perpendicular incidence angle, or relationship of the earth's surface to the direction of the sun's rays. (This situation is reversed for the southern hemisphere.) (Refer to Appendix A for uses of these angles in calculations for solar access.)

The *declination* of the sun is the angle measured by the direction of the sun at noon and the equator. It varies from month to month as the earth travels around the ecliptic. The angle changes from winter solstice (December 21) to spring equinox (March 21) by 23.5° and from spring equinox to summer solstice (June 21) by 23.5°.

The *latitude angle* is a measure between the equator and the north and south poles. At the equator it is 0°, and as it approaches either pole, it increases in value. The angle is expressed in degrees "north" in the northern hemisphere and degrees "south" in the southern hemisphere. Latitudes found in the United States range between 25 and 50° north.

The *hour angle* is a measure of the nominal time that it takes the sun to encircle the earth; this is approximately 360°. Beginning with true south, where the angle is 0°, the hour angle moves at a rate of 15° per hour. The east values are positive, and the west values are negative.

The *solar altitude angle* is a measure of the sun as it rises above the horizon and depends on three factors: the time of day, the day of the year, and the latitude. When the sun is on the horizon, the angle is 0°, and it can rise as high as 90°. At noon it is highest in the sky.

The *azimuth angle* is simply the angle formed by a line running due south and a line projected on a horizontal plane by the altitude angle. When the sun is due south, solar noon, the azimuth angle is 0°. To the east of south it is expressed as positive, and to the west it is negative.

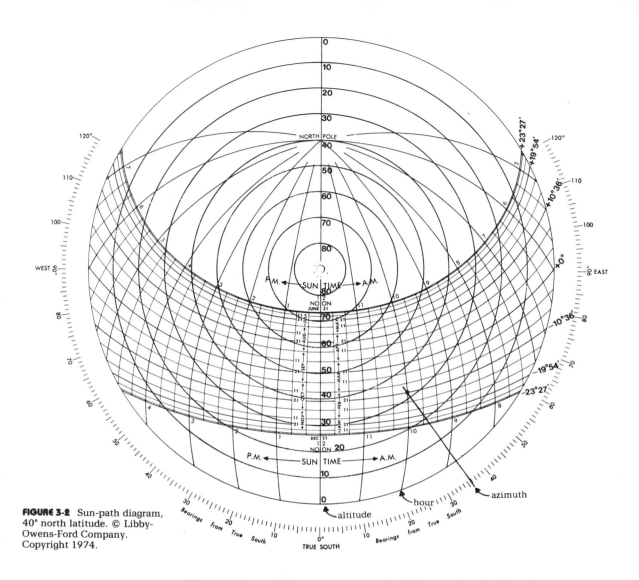

FIGURE 3-2 Sun-path diagram, 40° north latitude. © Libby-Owens-Ford Company. Copyright 1974.

The *profile angle* is a measurement of the projected altitude angle onto a vertical plane that is perpendicular to a surface. Usually the profile angle is projected onto a vertical plane located on the north-south axis.

The *incidence angle* is a measure of the sun's ray and a line perpendicular to a surface. The surface can be oriented in nearly any direction. The incidence angle is determined by the latitude, surface tilt angle, the declination, and hour angle.

The more perpendicular a solar collector is to the incoming solar radiation, the more energy it can collect. As the incidence angle approaches 0°, a collector surface reaches its greatest potential for collection. It is easy to see that higher latitudes receive lower amounts of solar energy, especially as the seasons change toward winter, and that the sun is less intense in the morning and afternoon.

Although the sun's position in the sky changes from minute to minute, hour to hour, and season to season, its position at any given time can be determined by the altitude and azimuth angles of a given latitude. For example, the sun can be located on January 21 at 9 a.m. at 40° north latitude by using the solar-position tables in Appendix A. The altitude angle is 16.8° and the azimuth angle is 44.0°. Understanding these angles and how they change becomes extremely important in determining shadow patterns and solar-access envelopes, which will be discussed later.

Sun-Path Diagrams

The movement of the sun can be plotted on hourly and seasonal bases by the use of sun-path diagrams. A *sun-path diagram* is simply a plot of the solar altitude, azimuth, declination, and hour angles in a polar-coordinate system. The diagram can be used to quickly locate the position of the sun for any time of the day, month, and year. There are many so called sun-path diagrams. For this book, the method developed by Irving F. Hand will be used. (Refer to Figure 3-2.)

The curved lines, which run horizontally across the sun-path diagram, represent the varying declination of the sun. At 40° north latitude, they are between +23°27′ and −23°27′. Refer to Table 3-1, which is used in conjunction with the sun-path diagram. It delineates the declination

TABLE 3-1
Declination of the Sun*

Declination	Approximate Dates
23°27′	June 21
20°30′	July 21
12°6′	August 21
0°	September 21
−10°42′	October 21
−19°54′	November 21
−23°27′	December 21
−19°54′	January 21
−10°36′	February 21
0°	March 21
11°54′	April 21
20°18′	May 21

*Adapted from ASHRAE.[2]

angles for the year. The curved lines running vertically calibrate the hours of the day. The concentric circles around the center of the diagram calibrate the altitude angles. The azimuth angles can be read along the circumference.

If, for example, you wanted to determine the altitude angle for February 21 at 10 a.m., you would first find the declination angle that corresponds with February 21. The value is −10°36′. Next, locate the hour angle line for 10 a.m. Where the hour angle line and the −10°36′ declination curve intersect is a point. From this point move around the nearest concentric

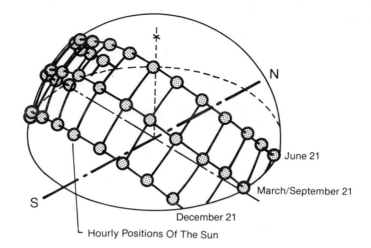

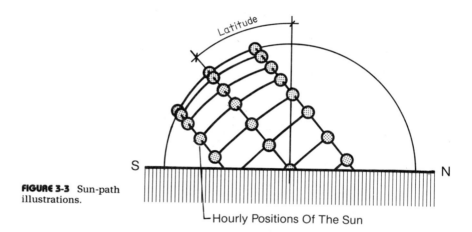

FIGURE 3-3 Sun-path
illustrations.

circle to the north-south line running vertically in the diagram. The value, which is approximately 33°, is the altitude angle for that time and date. The azimuth angle can be found by moving outward from the intersection of the declination curve and the hour line to the circumference. In this case, the azimuth angle is approximately 36°.

The sun-path diagram is a quick method for determining altitude and azimuth angles. It is also useful in visualizing the lines of movement of the sun throughout the year. Refer to Figure 3-3, which illustrates the sun path in isometric and elevation form. The altitude and azimuth angles are identified in Tables A-1 through A-5 at the end of Appendix A.

Solar Window

The *solar window* is an interval of time and space through which solar energy passes. It is generally regarded as the time for useful solar collection. The solar window is not determined by sunrise and sunset hours because of the high atmospheric absorption and the high incidence angles that occur at those times. The time is generally described by 9 a.m.

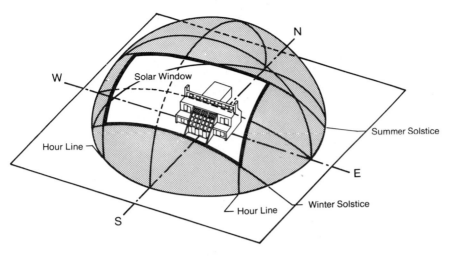

W

N

Solar Window

Summer Solstice

Hour Line

E

Hour Line

Winter Solstice

S

FIGURE 3-4 The solar window.

through 3 p.m. and by the winter and summer solstices. If you refer to Figure 3-4, you can identify the solar window by following the sun path during December 21 and June 21 and the hour angle lines for 9 a.m. and 3 p.m. The window varies in size depending upon the latitude and the selected hour lines. The upper and lower boundaries of the solar window correspond to different end uses. For example the upper boundary, which is determined by the June sun path, can describe solar access for domestic hot-water heating systems because they have a year-round use. On the other hand, the lower boundary, which is determined by the December December sun path, can describe solar access for both space heating and domestic hot-water heating systems for winter use.

The solar window can further be reduced in size through local obstructions, such as mountain formations, trees, or buildings. The opening area of the solar window is referred to as the sky space. It represents the optimum potential for solar energy collection. Critical to the solar energy design process is the relationship or juxtaposition of a passive or active solar energy system to the sky space. The more centered the orientation of the solar energy system to the sky space, the greater the potential gain. In other words, the solar collector should be positioned directly beneath the sky space and oriented to positions of the sun during the times of the year in which solar energy is required. Refer to Figure 3-4, which shows a solar house beneath the solar window.

To orient a building to the solar window, you first determine the center of the window, or true south. There are two methods. The first is to locate the north star (Polaris). There is a slight rotation of this star over a 24-hour period, but this can be overlooked. The other method is to use a compass. Make sure to adjust for magnetic variations. These can be found on the isogonic chart in Figure A-12. In the continental United States, these variations range from 0 to 24°. Once true south has been located, the overall width of the solar window can be determined by finding the

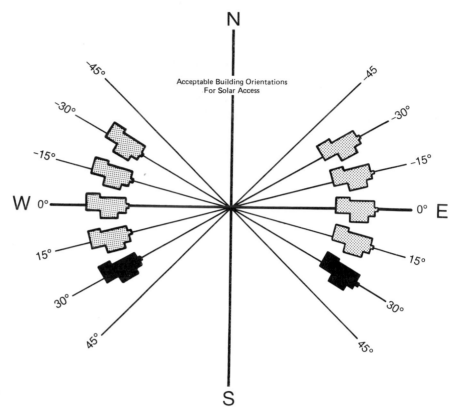

N

Acceptable Building Orientations
For Solar Access

-45°   -45

-30°   -30°

-15°   -15°

W 0°   0° E

15°   15°

30°   30°

45°   45°

S

FIGURE 3-5 Orientation
effects.

hour angles. The hour angles can generally be assumed to be symmetrical unless local physical or climatic conditions favor one side of true south to another.

The orientation of a building with respect to due south has an important effect upon its energy efficiency. Site conditions or even design considerations often dictate deviations from strict north-south orientations. A 10° variation to either the southeast or the southwest will not have a noticeable effect upon the performance of a solar energy system. A 20° variation will cause a small performance drop in efficiency. A 30° variation is considered the cutoff point for many state and federal tax credits for orientations of solar collectors. Beyond 30°, solar collection is considered to be uneconomical unless accompanied by collector insulation and window management. Figure 3-5 illustrates these orientation effects, and Figure 3-6 illustrates a site plan in which four multifamily housing units have slightly different orientations within these constraints.

SHADOW PATTERNS

Shadows cast upon a building or site can be either positive or negative depending upon the internal thermal and lighting needs. In summer in a warm climate, shade is welcome and the architectural form and land-

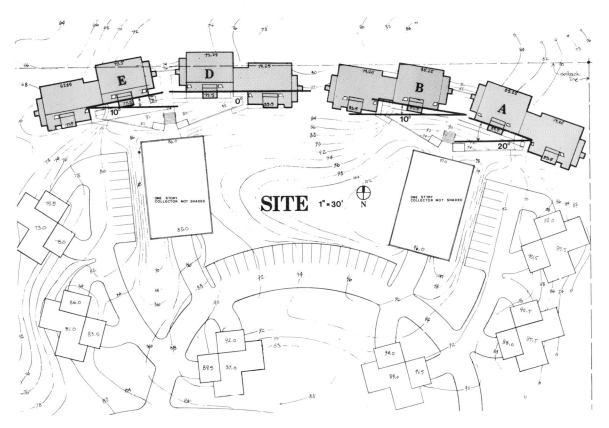

FIGURE 3-6 Example site plan, Briarwood condominium development. West Des Moines, Iowa. (Courtesy of Phillip Tabb, Roland Hower, and Alan Brown.)

scaping should help facilitate cooling. Obviously, in winter in a colder climate, as much sun as possible is desirable. In moderate climates, where the swing (spring and autumn) periods can be both warm and cold, a building should moderate natural heating and cooling. This can be accomplished through careful analysis of the shadow patterns, which occur both daily and seasonally, and an appropriate planning or design response to this analysis.

A shadow, in the broadest sense, is caused by the absence or exclusion of direct sunlight. The distinction between shadow and shade is somewhat academic but deserves mentioning. Both terms refer to the lack of sunlight. *Shade* is caused by the general absence of any sunlight on an area or object, while *shadow* is the direct result of an object obstructing the direct sunlight. An easy way to compare the two is through the use of an example. During a lunar eclipse of the sun, the dark side of the moon is in shade while at the same time the earth is in shadow. The moon in effect casts a shadow onto the surface of the earth. Both shade and shadow are related, but this section of the chapter will focus on shadows, specifically, those caused by building elements, single buildings, and multiple buildings. Understanding shadow patterns is an important step in planning for solar-access protection.

FIGURE 3-7 Shadows of building elements. This building is a HUD cycle 1 solar demonstration in Boulder, Colorado. (Courtesy of Phillip Tabb, Roland Hower, and Alan Brown.)

FIGURE 3-8 Interior photograph.

Building Elements

Many different building elements can be responsible for shading either passive or active solar collectors. Sometimes shading is planned, yet often it will occur when it is not wanted. Shading problems most often occur with more complex building designs where the building "footprint" has many "ins and outs." In designing for solar access at this level, an important thing to remember is the daylight hours of need with

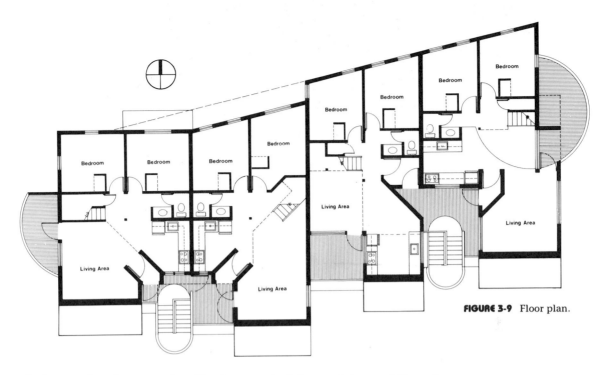

FIGURE 3-9 Floor plan.

their associated sun angles. Shadows cast at the periphery of the solar window are all right and account for minimal overall loss in a solar energy system's performance. Figures 3-7 and 3-8 illustrate shadows of building elements. In Figure 3-9, note the disposition of collector planes in order to avoid shading. Morning and afternoon shadows caused by south-projecting building forms miss the active collector arrays.

The exterior photograph in Figure 3-7 was taken around 9 a.m., when useful solar collection for the active collector arrays begins. The south facade is composed of several sloping building forms that are staggered in relation to one another along the north-south axis. Note that to avoid collector shading, the building forms are also positioned away from one another along the east-west axis. Early morning shadows normally extend in a northwesterly direction, and afternoon shadows are in a northeasterly direction. As the "jogging" of building forms becomes more exaggerated, shading is more apt to occur. The specific design and location of building elements such as building masses, wing walls, clerestories, dormers, elevator towers, collector arrays, chimneys, stacks, and flues need careful consideration to improve the solar energy collection potential of a building.

Single Buildings

In isolation, the shadow of a single building is unimportant. It should not affect its own collection or energy performance. When considering surrounding spaces or other buildings, it is important to know the shadow patterns during daily and seasonal collection periods. The shadows of

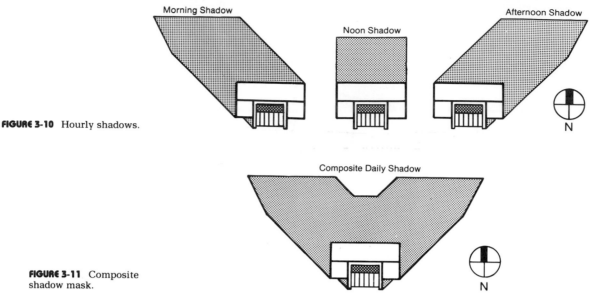

FIGURE 3-10 Hourly shadows.

FIGURE 3-11 Composite shadow mask.

single buildings are commonly delineated by a shadow mask. The *shadow mask* is a plot of shadows. December 21, which is the shortest day of the year, is usually chosen to generate the shadow mask. In the northern hemisphere, these are the longest shadows. The coldest day of the year can also be chosen to generate the shadow mask. This, of course, depends upon local weather conditions. The hours of the day used for the shadow mask are usually 9 a.m., 12 noon, and 3 p.m. Therefore, the mask is a composite of three shadows cast at these three times.

Figure 3-10 illustrates an hourly shadow sequence created by a single building at 40° north latitude on December 21. At 9 a.m. the altitude angle is 14.0° and the azimuth angle is 41.9°. At noon the altitude angle is 26.6° and the azimuth is 0°. At 3 p.m. the altitude angle is 14.0° and the azimuth angle is 41.9°. Figure 3-11 illustrates a composite shadow pattern for the same building. The shadow pattern seems to have two elongated wings in the northwest and northeast directions. The diagonal direction of these wings can cause shading problems especially where multiple buildings are concerned.

Topography can have an effect upon building shadows. Both the grade of the slope and the direction of the slope can change a shadow length. Where land contours slope upward in the direction of a shadow, the shadow will be foreshortened. The reverse is true for contours sloping downward; the shadows are longer. Buildings located on sites with a slope to the south will have shorter shadows and, therefore, can be placed closer together. On the other hand, buildings on north-sloping sites will have longer shadows and will need to be spaced farther apart. Figure 3-12 shows a shadow mask for a building located on a northeast-sloping site. Note how the morning shadow is lengthened while the afternoon

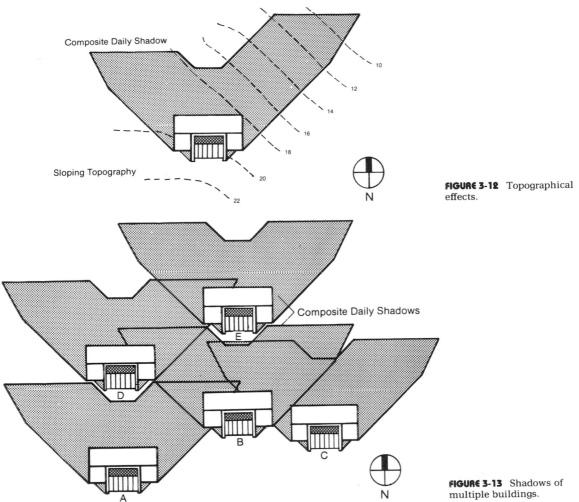

Composite Daily Shadow

Sloping Topography

10
12
14
16
18
20
22

N

FIGURE 3-12 Topographical effects.

Composite Daily Shadows

E

D

B

C

A

N

FIGURE 3-13 Shadows of multiple buildings.

shadow is similar to one on a flat surface. (Refer to Appendix A for methods of determining shadow patterns on varying topography.)

Multiple Buildings

Determining shadow patterns for multiple buildings can be more complicated than for single buildings. Often with projects of multiple buildings, there are topographical changes throughout the site, individual buildings that may have difficult orientations, and solar-access issues that need to be addressed. The situation can be aggravated by the need to achieve higher densities. Economic constraints can further buffet energy-oriented considerations. Shadow patterns of individual buildings have to be considered in the context of the entire site-planning process. The problems associated with the shadows of multiple buildings are described in the following examples.

Figure 3-13 illustrates a site plan with winter shadows of several buildings on a flat site. Buildings can generally be positioned along shadow lines to the northeast and northwest or behind the buildings to the north without suffering from shading. Building A, for example, casts a composite shadow as described before: hourly shadows at 9 a.m., 12 noon, and 3 p.m. on December 21. Building B is sited adjacent to building A along the afternoon shadow line. As long as building B is kept along the shadow line, it will be free of shading. Building D is sited behind building A. If it is moved any closer to building A, it will experience shading. Building C is located to the east of building A. Its shadow also limits the positioning of building E.

Another example of a multiple-building situation that could incur shadows or difficulty in getting solar access to every housing unit is the quadruplex or multiplex. A *quadruplex* is made of four attached housing units—usually with two adjacent sides. This affords opportunities for energy conservation. However, due to the nature of the quadruplex form, which is multidirectional, not all the housing units have an optimal orientation or exposure to the south. Two units are on the south, and two are on the north. The roof and the east, west, and south sides are kept free of shadows from neighboring buildings or landscaping elements. This helps ensure solar-access protection for all of the units even though the northern ones do not face south. Having sloped sites to the south can help with this housing form by allowing the northern units to be moved uphill where they can receive solar energy through clerestories.

The third example is the situation in which varying topography affects the placement of buildings. Buildings can be located closer together when topography is uphill because building shadows are shorter. And they should be located farther apart when topography is downhill because shadows are longer. When contours are parallel to the direction of a shadow, the shadow cast will have the same effect as one on level ground. (Refer to Appendix A for more detail on shadows and sloping topography.) If the sloping topography is relatively constant throughout the site, a shadow mask for a typical footprint can be made and positioned over the site plan like a template. This is a quick method for determining preliminary site-planning schemes.

SOLAR ACCESS

Several factors are involved in determining the proper levels of solar-access protection. A blanket or overgeneralized approach to solar-access planning is inflexible and naive. Buildings designed in climates with reduced sunlight, for example, are less inclined to need strict solar-access protection. Buildings located within higher-density zones may suffer severe shading from existing buildings. And buildings located in mountainous areas may need only partial solar-access protection depending upon the location of the mountains or hills. Solar access, therefore, needs to be examined within a local setting.

Climates that receive low levels of solar radiation during the heating season may not need extensive solar-access protection. These climates

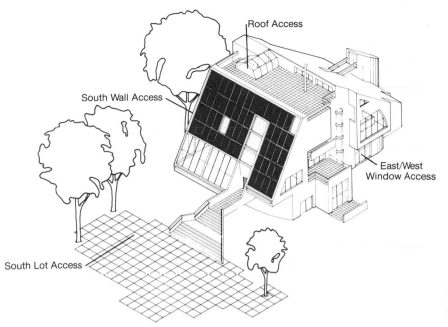

South Lot Access

Roof Access

South Wall Access

East/West
Window Access

FIGURE 3-14 Levels of solar access. From U.S. Department of Housing and Urban Development.[69,70]

are usually characterized by a great deal of diffused radiation. Typically, diffused radiation comes from all directions although it tends to be a little stronger to the south. Therefore, shadows from neighboring buildings or other obstructions may not be so critical. Climates with moderate to high levels of direct solar radiation can require more defined levels of solar-access protection.

Architectural or planning considerations can affect the degree to which solar access is needed. Development programs, budgets, and energy-design goals can vary from project to project. Where solar-access ordinances have been adopted, general levels of protection are ensured. This protection is usually limited to the context surrounding a particular project. However, ordinances have been adopted in only a few communities. Responsibility for solar-access design for the most part falls on the developers or designers themselves on a project-by-project basis. Therefore, specific levels of solar-access protection need to be determined for the context as well as for the buildings internal to the project.

Levels of Solar-Access Protection

The levels of solar access are defined by the portions of the building envelope that are normally exposed to sunlight, that is, the roof and the east, south, and west walls. Solar access can also be given to other exterior spaces that may need it. For the purpose of simplification, there are four levels of solar access. They are roof access, south wall access, east-west wall access, and south lot access. Figure 3-14 indicates these four levels. There is generally a hierarchy of protection. It moves from the level in need of greatest protection, the roof, to the next level, the south wall, to the next level, the east and west walls, and finally to the level in need of least protection, the south lot. As the solar-access protection becomes more strict, fewer levels are used. Descriptions of these various levels follow in order of importance.

Roof access is the most conservative level of solar access. It obviously occurs at the highest elevation of the building, which is least likely to be in shade. This level of access is appropriate to higher-density developments. It protects a variety of rooftop solar collection methods, including active collector arrays, clerestories, skylights, roof monitors, and even greenhouses. If only roof access is achievable, care should be given to the thermal coupling of the collection method and the spaces for which it serves. Rooftop solar energy systems should also be designed so that shading does not occur. Watch out for elevator towers, conventional mechanical equipment, and other building elements.

South wall access makes possible common applications of passive heating: direct-gain, thermal-wall, water-wall, and sunspace systems. Both vertical and tilted collection surfaces can be protected on the south. South wall access gives good protection for buildings employing sun-tempering systems, incremental passive systems, full passive systems, and a variety of active systems. South wall access normally includes roof access, which allows for a certain degree of design flexibility. Care should be taken between collection devices and other building elements (stairs, balconies, overhangs, etc.).

East-west wall access is becoming more important with buildings located in climates with lower levels of sunlight and with buildings that have more complicated thermal zoning. The added protection to the east and west gives more access potential. Single-family detached and duplex housing types, for example, can benefit greatly from this kind of access. In developments or neighborhoods of higher densities, it is difficult to maintain a consistent level of east-west solar access. This level of access may be beneficial, but it may need to be accomplished on a "do what you can" basis.

South lot access protects ground-level exterior spaces to the south, southeast, or southwest of a building. Certain outdoor spaces require direct sunlight, such as patios, courtyards, swimming pools, and gardens. Entries or detached collection devices may need south-lot protection. Even ground-level collection devices adjacent to south facades may need further protection. This level of solar access may differ from others in its need for year-round protection. Outdoor activities most often occur during spring, summer, and autumn months when weather is milder. Sun angles are higher at these times, making south lot protection easier to accomplish.

The levels of solar access explained here are generally associated with single buildings; therefore, they deal with the internal needs for protection. In a planned unit development, for example, consideration of these levels would probably mean appropriate site planning to ensure solar-access protection for the individual buildings of the project. Protection

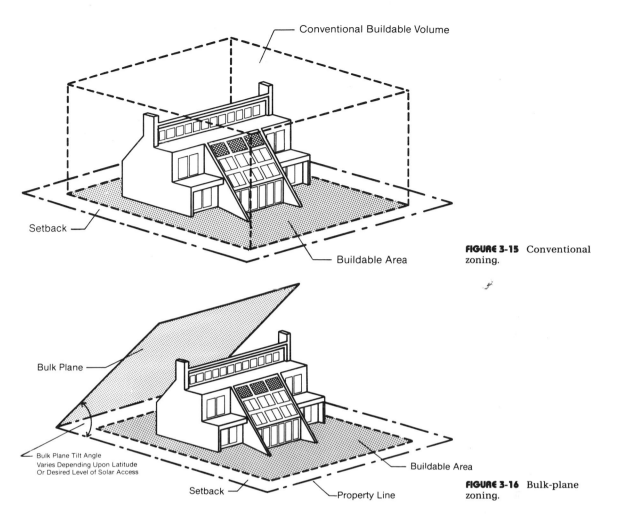

FIGURE 3-15 Conventional zoning.

FIGURE 3-16 Bulk-plane zoning.

should also be considered for neighboring buildings that may surround a particular project site. Context protection, as it may be called, is addressed with the concept of solar zoning.

Solar Zoning

Solar zoning is a zoning concept that defines the volume, size, and shape of space that encloses a building site and limits shadowing to neighboring buildings. It aims to provide solar-access protection (context protection not necessarily internal protection). There are several methods of solar zoning.

The first method is called the *bulk-plane* method. Essentially, it uses a graduated height from the north to the south of a building site. Either a sloped plane or ziggurat (stepped mass) is used. Refer to Figures 3-15 and 3-16, which illustrate a conventional zoning envelope and the bulk-plane zone. The bulk-plane method cuts down on usable (based upon conventional zoning) building volume. The northern portion of the buildable

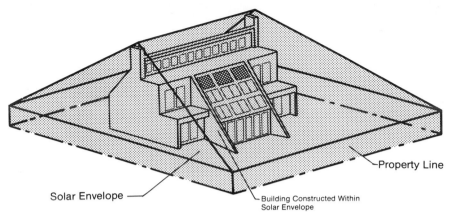

FIGURE 3-17 Solar envelope.

Solar Envelope

Property Line

Building Constructed Within
Solar Envelope

volume of the site is reduced in height while the southern portion is increased in height. If this method of solar-access control does not compensate for the reduced volume of buildable space, it may receive severe criticism from developers and builders. The virtue of this method is its simplicity. With this method, you use the seasonal sun angles only.

The second method, the *solar-envelope* method, was created by Professor Knowles at the University of Southern California. According to him, "The idea of the solar envelope moves beyond concepts of conventional zoning because the latter traditionally delimit space in static terms. But the space within the solar envelope, and hence the implications for design and development, is defined dynamically by the rhythms of solar movement at daily and seasonal intervals."[32] The method is primarily the result of envelope generation through the superimposition of daily and yearly sun angles. Figure 3-17 illustrates a generalized solar envelope that responds to the winter demand for solar access. The figure also illustrates a generalized envelope that responds to daily and seasonal demands. A complete solar envelope is formed by superimposing planes formed by the daily and seasonal sun angles upon a specific site. The solar envelope is a good solar-access tool as long as it remains fairly simple. The more complicated the solar-envelope form, the more difficult the legal process of defining that form. Because the envelope changes from site to site, generating and maintaining this zoning form may not be desirable for many planning agencies. This solar-access method is site-specific.

The third method is called the *solar-fence* method. It was adopted by the City of Boulder, Colorado. The method limits the perimeter around a given property by means of a solar fence, or imaginary wall. Figure 3-18 illustrates this concept. Buildings within the property are not allowed to cast shadows above the height of the given solar fence. Three solar-access zones are related to existing land-use zones. Each zone has a different solar-fence requirement. The first zone is for low-density residential areas. The solar-fence height is 15 feet. The second zone is for higher-density residential areas, and the solar-fence height is 25 feet. The third

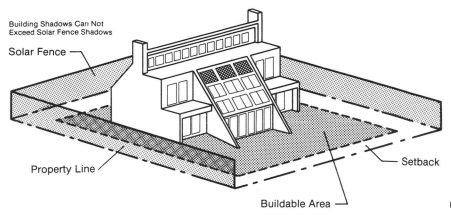

Building Shadows Can Not
Exceed Solar Fence Shadows

Solar Fence

Property Line

Buildable Area

Setback

FIGURE 3-18 Solar fence.

zone is commercial, and no solar fence is required. This is a relatively simple solar-zoning form.

The solar fence is placed around the perimeter of the site. Shadows caused by buildings cannot exceed this height. The shadows are determined at 10 a.m., 12 noon, and 2 p.m. This approach assumes that normal front-yard setbacks in conjunction with the solar fence will provide adequate solar-access protection. This approach, as described for the City of Boulder, Colorado, is conservative, especially for the higher-density residential zones where protection is generally provided for roof access only. Refer to Appendix B for more detail.

DENSITY

According to the *Residential Development Handbook* published by the Urban Land Institute, "Development density has been and continues to be a much over-used index among developers, and particularly among some public officials who view density as the sole measure of a project's acceptability."[72] Residential density is determined by the number of people or the number of dwelling units for a particular project site. It is based on either gross land area or net land area. Gross density is the total land area of a site, including streets, public parking, sidewalks, playgrounds, and other nonresidential areas. Net density is the land area devoted exclusively to buildings and accessory uses on the same site.

The density of a development is not a measure of its quality; that is, lower densities are not better than higher densities, or vice versa. For example, a low-density development with large lot areas per dwelling unit affords opportunities for ground-level living spaces, good solar access, and noise and privacy protection, to name a few things. However, these do not ensure quality. Similarly, a high-density development affords the opportunity for easy access to many support facilities, recreational opportunities, etc. These also do not ensure quality. Density is merely a way of classifying a particular development in terms of numbers of units in relation to open space.

Density is important in relation to the question of solar access. Generally, it is easier to achieve solar access in low-density situations. Conversely, higher densities usually result in solar-access problems unless solar energy has been carefully considered. The need to plan for solar access tends to intensify with increased densities. The relationship between density and solar access is one of the greatest challenges to the large-scale utilization of solar energy in buildings, especially within the urban context.

There are two areas of concern for solar access in relation to achieving higher densities. The first is *context protection.* This refers to ensuring solar access to buildings that may exist outside of a development. Greater levels of solar access can be given to building types that may require it. On the other hand, lower levels can be given to building types that do not necessarily need it. Unless specified in a solar ordinance, the responsibility for determining these various levels of solar access for adjacent sites lies solely with the designer or developer.

The second area of concern is *internal protection.* This refers to ensuring solar access to buildings and/or housing units found within a development project. Planning for internal protection requires more skill and precision as densities increase. Similarly, the levels of solar access will generally relax with increased density. This is usually due to the economics and the practicality of the planning and architectural process. Care should be taken in building location, massing, and location of solar collection devices. Internal protection is examined with the following three densities.

Low Density

Low-density development, planned with a modicum of sensitivity to the movement of the sun and to the shadows that may result, should be capable of achieving all of the levels of solar access: roof, south wall, east-west wall, and south lot access. Figure 3-19 is a model photograph illustrating a low-density development of four units per acre. The elongated site is 1 acre. For the purpose of discussion, the site is flat and oriented south. It is located at 40° north latitude. The design provides for solar access of no less than 6 hours' duration on December 21, the design day. For most designs at this density, 6 hours should be easily achieved. The scheme has four single-family housing units on four separate parcels of land. The average footprint area of a unit is 1500 square feet, and the houses are typically one or two stories in height. The size of a single plot is approximately 10,000 square feet, and the total development size is 1 acre.

Two of the houses front to the south, and two front to the north. An alley separates the north and south parcels. Based upon this arrangement, sunny or southern open spaces tend to serve different functions. For the units oriented to the south, the southern spaces are more formal; the covered porches make the entry more elaborate; orientation to the street is more pronounced; and landscaping tends to be more formal, with flower gardens. For the units oriented to the north, the southern spaces

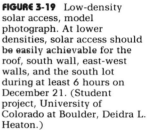

FIGURE 3-19 Low-density solar access, model photograph. At lower densities, solar access should be easily achievable for the roof, south wall, east-west walls, and the south lot during at least 6 hours on December 21. (Student project, University of Colorado at Boulder, Deidra L. Heaton.)

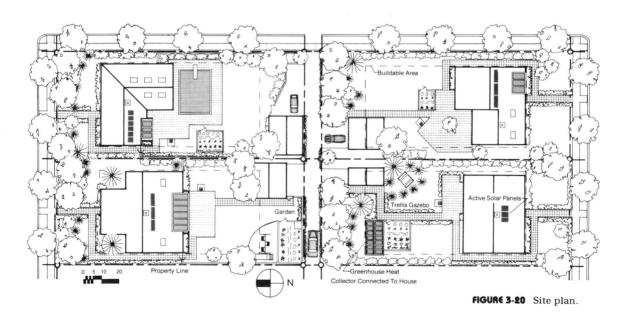

FIGURE 3-20 Site plan.

are more informal; southern exterior spaces tend to be patiolike, as an extension of the indoor activities; activities are oriented toward recreation and family events; and landscaping is more personal. Figures 3-19 through 3-22 illustrate this low-density example.

The solar-access analysis clearly illustrates that solar access is possible

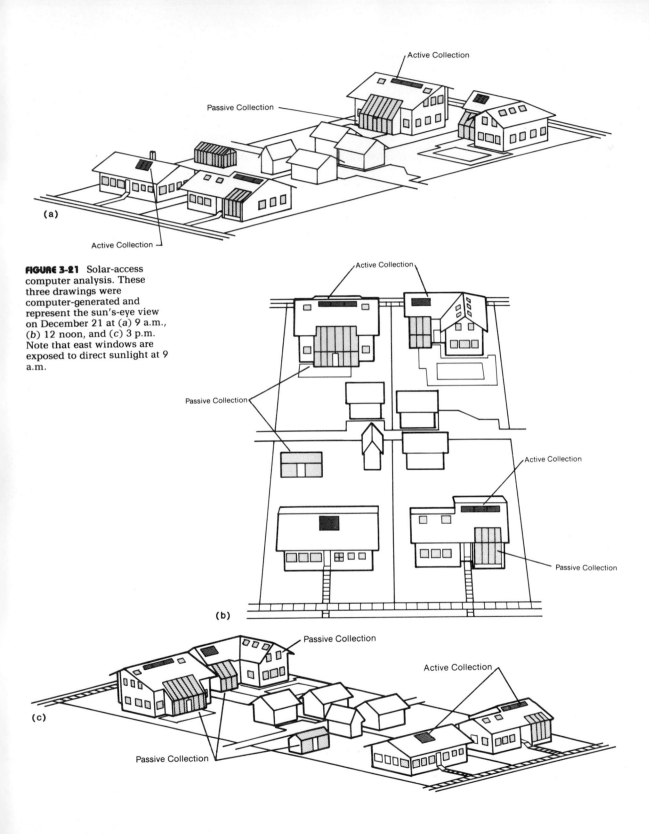

FIGURE 3-21 Solar-access computer analysis. These three drawings were computer-generated and represent the sun's-eye view on December 21 at (a) 9 a.m., (b) 12 noon, and (c) 3 p.m. Note that east windows are exposed to direct sunlight at 9 a.m.

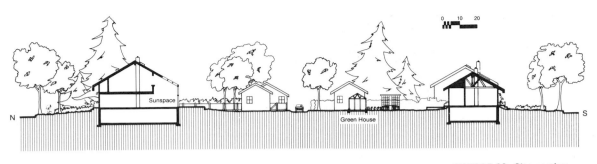

FIGURE 3-22 Site section.

during 6 hours of the day. Note the east windows, which are clearly exposed to the morning sun. This analysis was accomplished with the help of a computer-aided program called Sunray. The program produces three perspectives of the project from the sun's-eye view (altitude and azimuth) for 9 a.m., 12 noon, and 3 p.m. on December 21. The dark areas indicate active solar collectors for domestic hot-water heating, and the light areas indicate passive collectors.

The roofs of the housing units have adequate solar access as do the south walls. The southern portions of the east and west walls also have access. Note that the area directly south of each of the houses is free of shadow, thus providing south lot access. Clearly this example illustrates the ease of providing solar access at this level of development density. Having fewer units per acre, obviously, will provide even greater flexibility in siting and design.

Medium Density

Solar access for medium-density development, of around twenty-five units per acre, is a little more difficult to achieve. Both context and internal protection are important to preserve. Here, only roof and south wall access are consistently protected throughout a project with this level of density. Figures 3-23 through 3-27 illustrate a development on a 1-acre site with roof and south-wall access for a medium-density project. The size, or area, of the site is the same as the low-density and high-density examples. The site is assumed to be flat, south-oriented, and free of any unusual conditions. The design provides for solar access of 5 hours' duration on December 21.

The design comprises two fairly low-rise buildings (four stories each). One is located on the southern portion of the site, and one to the north. Each residential unit has 1000 square feet. A 12-foot solar fence has been assumed to ensure context protection. As a result, the building locations and heights vary according to their shadows in the morning, at noon, and in the afternoon. The maximum number of stories from the ground is three. South-facing windows, balconies, and sunspaces are positioned or massed in such a way as to be free of internally produced shadows. On the southern portion of the roof are three active collector arrays for

FIGURE 3-23 Medium-density solar access, model photograph. At medium densities, solar access should be achievable for roof and south wall for 5 hours on December 21. Shading is likely to occur on east and west walls and the south lot. (Student project, University of Colorado at Boulder, Stephen Glascock.)

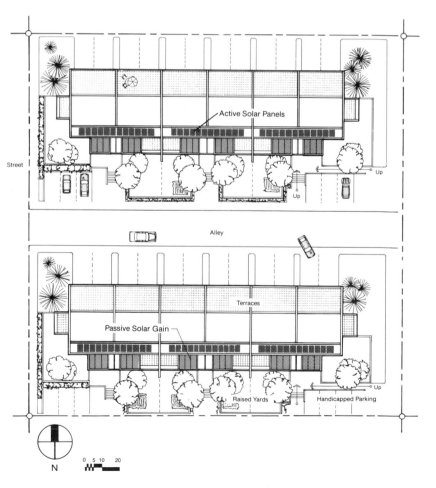

FIGURE 3-24 Site plan.

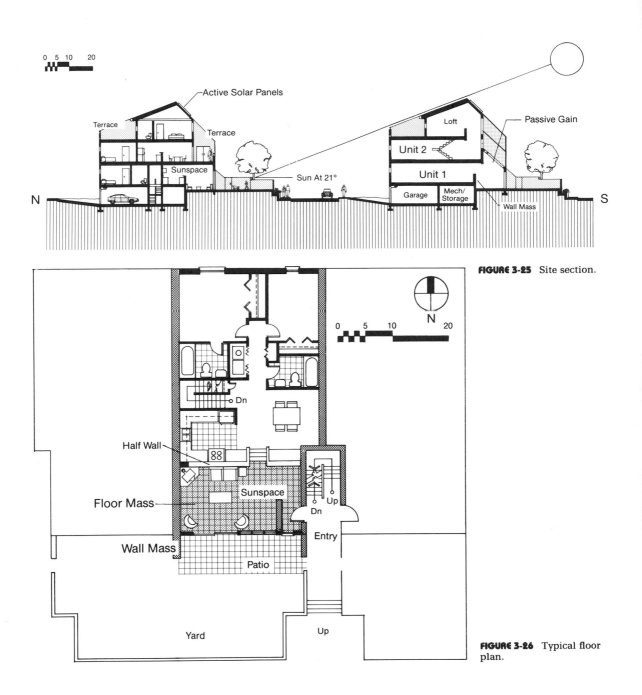

FIGURE 3-25 Site section.

FIGURE 3-26 Typical floor plan.

domestic hot-water heating. As can be seen in Figure 3-27, there is no shading from 9:30 a.m. to 2:30 p.m. on December 21.

High Density

Adequate solar access for high-density developments, of about fifty or more units per acre, is obviously much more difficult to achieve. Quite

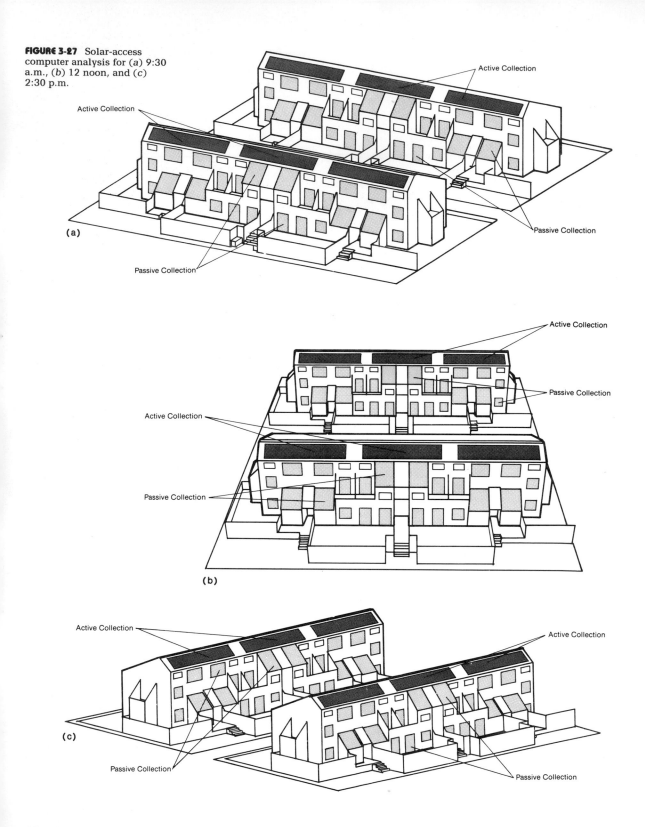

FIGURE 3-27 Solar-access computer analysis for (*a*) 9:30 a.m., (*b*) 12 noon, and (*c*) 2:30 p.m.

Active Collection

Active Collection

Passive Collection

Passive Collection

(a)

Active Collection

Passive Collection

Active Collection

Passive Collection

(b)

Active Collection

Active Collection

Passive Collection

Passive Collection

(c)

FIGURE 3-28 High-density solar access, model photograph. At high densities, solar access is more difficult. Roof and south-wall access are achievable for 4 hours on December 21. Care should be taken in order to prevent internal shading. (Student project, University of Colorado at Boulder, David Horsley.)

often a project of this density is located in or adjacent to other high-density developments, where shading from other buildings is bound to occur. Roof access and as much south wall access as possible are reasonable goals for each housing unit. This, of course, depends upon the specific design. For example, double-loaded corridors with housing units on both sides of a corridor will suffer solar-access inequalities. Single-loaded schemes with the housing units facing south afford good solar access. Solar collection devices, both active and passive, may be singled out for protection at this density. South lot and east-west wall access are difficult to achieve on a consistent basis and, therefore, are compromised. The design should provide for solar access of at least 4 hours' duration on December 21. However, longer hours are possible. In the example illustrated in Figures 3-28 through 3-32, there is no severe shading from adjacent buildings outside of the site. For context protection, a 25-foot solar fence has been assumed.

The design is located on a 1-acre site and has forty-nine 1000-square-foot housing units that are all massed into one structure. The building is terraced, and all units have a southern orientation with private spaces, terraces, or balconies facing south. The public activities and spaces are placed on the north and ground floor. Amenities include a day-care center, an indoor multipurpose space with recreational facilities, a shared laundry and kitchen, a playground, and swimming pool. On the roof are located two community greenhouses for year-round growing and plant care. Because of the numbers of units and the overall height of the build-

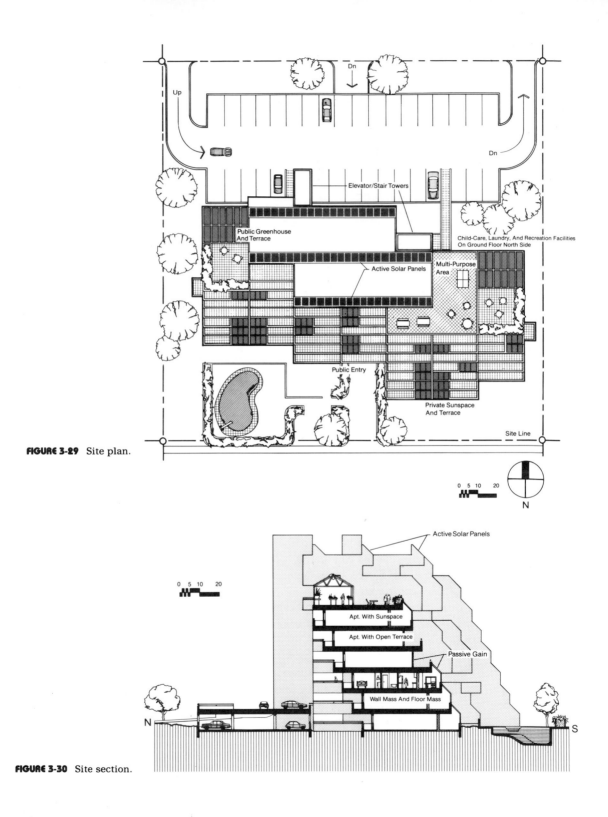

FIGURE 3-29 Site plan.

Dn

Up

Dn

Elevator/Stair Towers

Public Greenhouse
And Terrace

Child-Care, Laundry, And Recreation Facilities
On Ground Floor North Side

Active Solar Panels

Multi-Purpose
Area

Public Entry

Private Sunspace
And Terrace

Site Line

0 5 10 20

N

FIGURE 3-30 Site section.

Active Solar Panels

0 5 10 20

Apt. With Sunspace

Apt. With Open Terrace

Passive Gain

Wall Mass And Floor Mass

N

S

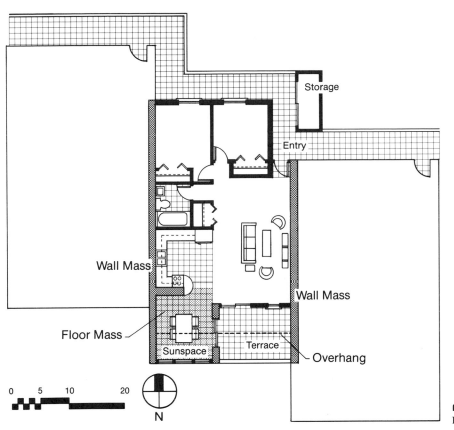

Storage

Entry

Wall Mass

Wall Mass

Floor Mass

Sunspace

Terrace

Overhang

0   5   10   20

N

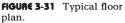

FIGURE 3-31 Typical floor plan.

ing, elevators are necessary. Parking occurs in a two-story structure located on the northern portion of the site. The energy systems include sunspaces or direct-gain passive systems for each individual unit and three rooftop active collector arrays for domestic hot-water heating.

Circulation occurs on the north. There are two vertical circulation towers with stairs and elevators. Enclosed or open corridors connect the towers to entries. Entries to the housing units also occur on the north. The zigzagging of the plan allows for more private entries. The floor plans are organized into three interior zones and two exterior zones. Beginning with the north is the exterior entry zone with storage. Inside the unit is the bedroom zone, the living-room zone, and the sunspace. To the extreme south of the unit is a terrace or balcony. Figure 3-31 illustrates the thermal-mass placement on the southern portion of the plan. Figure 3-32 shows the sun's-eye view at 10 a.m., 12 noon, and 2 p.m. on December 21. There is no appreciable shading. Note that on the back active collector array there is a little shading caused by the elevator tower. This should be avoided.

As densities increase, designing for solar access is more difficult. There

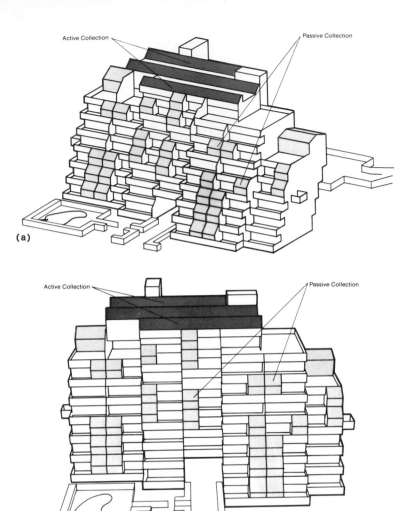

Active Collection

Passive Collection

(a)

Active Collection

Passive Collection

(b)

are ways to reduce the difficulties. The most obvious way is to reduce the size requirements for the various areas, both indoor and outdoor. This also includes reducing building volumes if possible. Housing units, for example, can be programed to have a lower overall area. Private open spaces can be reduced and replaced with public spaces. The degree to which a program is reduced most certainly needs careful scrutiny. If the units are too small, other problems are sure to surface. Then the units may be more difficult to sell or market, and they may stress internal function. The relationship between increased density and solar access is very important and may stimulate interesting change in planning practice.

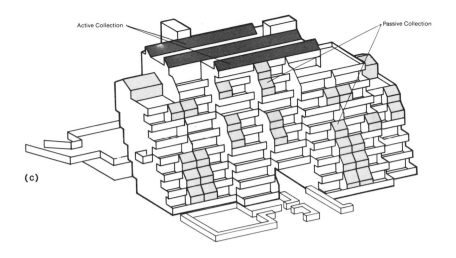

Active Collection

Passive Collection

(c)

SOLAR-ACCESS DESIGN PROCESS _____

A precise checklist or method of solar access design is difficult to determine because of the many complex situations that commonly occur. Building programs, specific sites, economics, and user requirements all play an important role in the creation of planning and architectural form, and they vary from project to project. A detailed "how-to" menu could not respond to these varying conditions. More appropriate, perhaps, is an attempt to present a more general list of "what to do" in the event that solar-access protection is desired. The following solar-access design process is divided into four steps: determination of the solar window, solar-access strategies, solar energy systems strategies, and solar-access analysis.

1. Determination of the solar window
   a. Determine the solar-access design day of the year (typically December 21 or the coldest day).
   b. Determine the desired hourly duration of sunshine (typically 4 to 8 hours during the design day).
   c. Identify governing altitude and azimuth angles associated with the duration of sunshine to determine shadow and shading patterns.
   d. Determine true south. You can do this by observing the north star or by using a compass. Check the isogonic chart for magnetic variations (refer to Appendix A, Figure A-12).
   e. Establish range of building orientations (determine off-axis orientation adjustments by revising shadow patterns) and topographical effects.
2. Solar-access strategies
   a. Identify any local site obstructions to the solar window (trees, buildings, mountain formations, etc.).

    b. Establish a method for determining context protection (bulk-plane, solar-envelope, solar-fence, etc.).

    c. Determine desired density and associated levels of solar access (roof, south wall, east-west wall, south lot, or other).

3. Solar energy systems strategies

    a. Determine preliminary energy balance (identifying emphasis on space heating, cooling, and/or conservation, etc.). Identify heat load, energy-conservation concepts, and methods of solar collection.

    b. Determine preliminary energy design strategies with location of south-facing windows and collectors (determine preliminary active and passive collector areas and thermal-mass volumes) and identify unit zoning concepts (east-west zoning, north-south zoning, etc.).

    c. Identify relationships of solar collectors to mass and determine solar-access requirements for thermal mass.

4. Solar-access analysis

    a. Ensure that context protection has been achieved (compare building forms with solar envelope or building shadows with solar fence).

    b. Ensure that internal protection has been achieved. Construct building-shadow patterns based upon the above items (look for shading on internal thermal mass, south-facing windows, and other solar collectors).

SUMMARY

Solar access, in concept, may appear to be simple and straightforward. But applied to the present development process, it is a more difficult proposition. There is great economic pressure in the building industry, and solar access may be seen as another negative limitation to the delicate manifestation process. As a consequence, guardianship of solar access must come from several sources. First, it must come from the community, in the form of zoning-ordinance reform. Solar-access provisions can be initiated and enforced. However, they tend to be conservative and too general and usually only provide context protection. Second, it must come from developers, in the form of covenant protection. This method usually provides only internal project protection. Covenants, considered alone, will tend to give spotty or unequal protection from project to project. Even with its limitations, this method is seen as the most practical solution. Third, it must come from design professionals, in the form of solar-access planning and design. This method, too, may be inconsistent and difficult to enforce.

    The need to plan for solar access is without question if we are to incorporate solar energy systems into both the future and existing stock of buildings. Solar energy, as seen as a partial replacement for conventional fuels, has economic strength to encourage change. With greater solar-access protection, another issue arises. That is, the issue of the quality of life. To quote Ralph Knowles: "The sun is fundamental to all life. It is the source of our vision, our warmth, our energy, and the rhythm of our lives. Its movement informs our perceptions of time and space and our scale in the universe."[32] Solar access, in this broader sense, transcends the practical considerations of just access to energy and suggests new territory that touches basic human rights. Solar access becomes open-ended to include creative uses of solar energy that have not yet been conceived.

SHELTER DESIGN 4

*It's getting cold again over here—and always
when it does I start thinking about how to
warm up architecture, how to make it lodge
round us. After all, people buy clothes and
shoes the right size and know when the fit feels
good!* ALDO VAN EYCK*

The housing unit, whether attached or detached, is the basic element
of residential settlement design. It is necessary to understand energy
planning determinants in relation to this scale of individual and family
needs. Therefore, focus on the nature of shelter design should enhance
the energy planning process and give a stronger connection to these basic
needs. It is the purpose of this chapter to relate the housing unit, with its
energy needs, to the broader problems of residential solar energy
planning.

The decentralized energy-use patterns associated with millions of individual homes have led to high energy-consumption rates. Substantial
energy savings can be achieved through improved shelter design. The
evolution of the shelter form can be responsive to the external effects of
climate as well as the internal energy needs and use patterns. The principal functions of the housing unit are presented in this chapter as they
form a basis for shelter design. Both conventional and solar energy considerations are related to *energy support systems, shelter design,* and the
scale of *small-parcel development.* This look at shelter design is the first
in a sequence that expands in scope and scale in the remaining chapters
from shelter to residential settlement.

*Quoted in Smithson.[54]

FIGURE 4-1 Shelter design examples: (*a*) Osheroff House model (courtesy of Phillip Tabb Architects), (*b*) the Acorn House, and (*c*) the Chubb House (courtesy of Phillip Tabb, Roland Hower, and Alan Brown). *(Continued on next page)*

(b)

ENERGY SUPPORT SYSTEMS

There is no doubt that the present energy support systems help make housing more livable, that is to say, more convenient. Central heating and air-conditioning, hot-water heating, and the myriad of electric devices aid in the day-to-day process of living. Many of the uses of energy are critical for the support of life, such as providing protection in extreme climates. The various forms of energy not only support the more mun-

(c)

dane functions of life but provide a great deal of leisure activity and entertainment as well. Our kitchens and living rooms are literally filled with equipment, machines, devices, and gadgets—ovens, ranges, refrigerators, toasters, blenders, vegetable choppers, electric knives, disposals, clocks, timers, can openers, telephones, radios, televisions, stereos, home computers, and games and toys for our children, to name a few. What would contemporary life be without all of these modern conveniences?

We know that the primary energy needs in the home are space heating, air-conditioning (depending upon climate), hot-water heating, food preparation, and clothes washing and drying. Therefore, a look at these basic end uses and the types of energy support systems will give a better understanding of the pressures placed on the decision makers at the architectural and planning levels. Figure 4-2 illustrates two energy-end-use pie diagrams. They were prepared by the Colorado Energy Research Institute. They represent electricity and gas consumption for the residential sector of the state of Colorado, a temperate climate region. Electricity consumption seems to be fairly evenly distributed among uses, while gas consumption is dominated by space heating.

In this section, an attempt will be made to focus on those functions in which energy conservation and solar energy systems may be applied. We

FIGURE 4-1 (Cont.) Shelter design examples: (*d*) the Wolff House and (*e*) a prairie house.

(d)

(e)

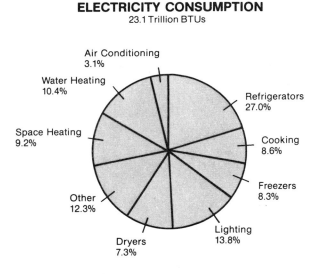

ELECTRICITY CONSUMPTION
23.1 Trillion BTUs

Air Conditioning
3.1%

Water Heating
10.4%

Space Heating
9.2%

Other
12.3%

Dryers
7.3%

Lighting
13.8%

Freezers
8.3%

Cooking
8.6%

Refrigerators
27.0%

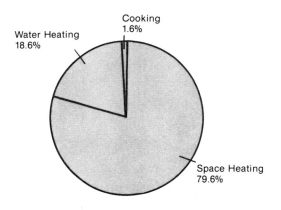

NATURAL GAS CONSUMPTION
127.4 Trillion BTUs

Cooking
1.6%

Water Heating
18.6%

Space Heating
79.6%

FIGURE 4-2 Residential energy consumption. (Courtesy of the Colorado Energy Research Institute.)

will look at the hierarchy of home energy needs and at the kitchen, which is most often the heart of the home and has a diversity of energy needs and support systems.

Home Energy Needs

Energy sources most commonly used for space heating are gas, electricity, propane, fuel oil, and solar energy. These energy sources usually heat either air or water for distribution throughout the home. At the present, gas is the lowest-cost fuel, excluding, for the moment, solar energy. In areas of the United States where gas is not as available, electricity and fuel oil are used. In remote areas propane is typically used—often with wood for wood-burning stoves. The types of heating systems vary from region to region but generally fall into several categories. According to a study prepared by the National Association of Home Builders (NAHB) Research Corporation, the mix of fuel and heating systems varies throughout the United States. For example, look at the following four cities in 1979.[45a]

1. *Madison, Wisconsin (cool climate).* 72 percent of the dwellings use gas, and 82 percent hot-air heating systems.
2. *Phoenix, Arizona (hot-arid climate).* 99 percent of the dwellings use electricity, and 87 percent heat pumps.
3. *Houston, Texas (hot-humid climate).* 55 percent of the dwellings use gas, 45 percent electricity, and 97 percent central air-conditioning systems.
4. *Denver, Colorado (temperate climate).* 86 percent of the dwellings use gas, and 91 percent hot-air heating systems.

Solar-assisted heating systems interface with these conventional systems or, in the case of many passive systems, function independently.

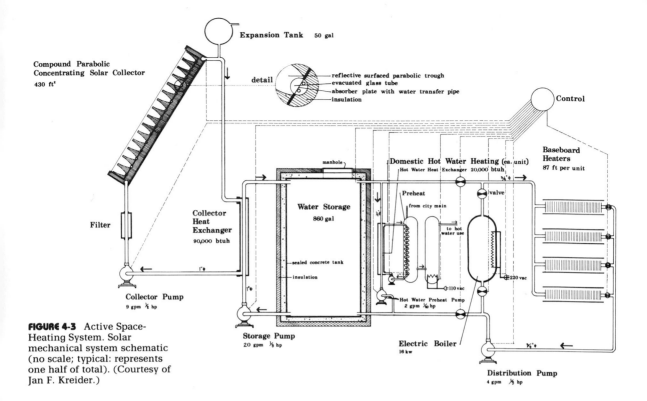

FIGURE 4-3 Active Space-Heating System. Solar mechanical system schematic (no scale; typical: represents one half of total). (Courtesy of Jan F. Kreider.)

Figure 4-3 illustrates a typical solar-assisted space-heating system using hot water (hot air can also be used). And Figure 4-4 illustrates a hybrid heating system that is both active and passive. Solar coupling to conventional hot-air systems usually occurs in the return-air distribution ducts, where the solar heat is treated as preheated air. Solar coupling to conventional hot-water systems usually occurs in the form of a heat exchange to the return loop of a baseboard distribution system. Electrical resistance heating has no direct connection to solar heating systems.

Hot-water heating systems use either gas or electricity. The interface of conventional systems and solar energy systems is quite simple. In the solar collector, water is heated directly by the sun. It is transported to storage, usually in the basement or mechanical room of the house, where it passes through a heat exchanger and returns back to the collector. The heat exchanger transfers heat to a storage tank, typically holding 80 gallons. When there is a call for hot water, it is drawn from the solar storage tank and distributed for use—typically for dish washing, clothes washing, showers, etc. When all the heat is taken from the solar storage, the conventional hot-water heating system is automatically turned on using either gas or electricity.

Electric networks are common to nearly every home in the United States. They are usually located in stud walls, floors, and ceilings and

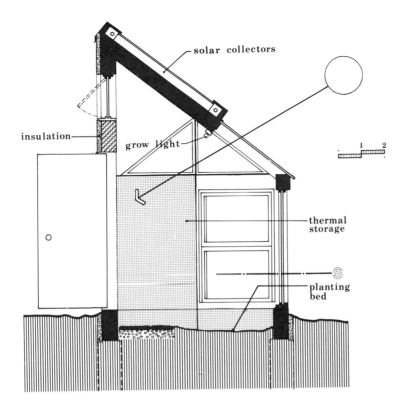

solar collectors

insulation

grow light

thermal storage

planting bed

1 2

FIGURE 4-4 Greenhouse section illustrating hybrid solar energy system.

provide electricity for lighting and other electrical needs. Most building codes demand that electric outlets be located every 12 feet around the perimeter of each occupied room in the house. This allows for a great deal of flexibility for the many electric fixtures of the home. The complexity of the electric circuiting is directly related to the complexity of the design. However, the impact of electric networking in a house is minimal, since the systems are concealed. The greatest energy saving comes from improved engineering of electric devices. Consumption of electricity could be dramatically reduced if all appliances were energy-efficient. In a typical home using 1000 kilowatt-hours per month, consumption can be reduced to between 300 and 500 kilowatt-hours per month.

The Kitchen

For the purpose of discussion here, a *kitchen* is defined as a service center for the house and therefore includes the utility and mechanical functions. A lot of time is spent in the kitchen. In a family context, the kitchen can be used extensively three times a day, with many other intermittent visits for snacks, etc. Often the kitchen is a social center of the home. As a result, a great deal of energy is used. If you analyze energy

TABLE 4-1
Common Kitchen Energy End Uses

	kWh/month
Range with oven	98
Dishwasher	30
Refrigerator with freezer (14 ft³)	95
Disposal	3
Toaster	3
Mixer	1
Coffee maker	8
Blender	1
Microwave oven	16
Lighting (service center only)	30
Clothes dryer	83
Automatic washing machine	9
Iron	12
Sewing machine	1
Hot-water heater	352*
Total	742

*Because of seasonal temperature differences in cold-water supply, hot-water heaters will use more energy in winter months than in summer months.

end-use diagrams for residential energy consumption in your locality, you will probably find that over 30 percent of household energy is consumed in this service center. A great deal of service hot water is used for washing and cleaning, and a large amount of electricity is used for lighting, food preparation, and refrigeration. Many electric devices are typically used in the kitchen. Table 4-1 lists many of these devices and pieces of equipment and gives an approximate energy consumption for each.

One of the key concepts for kitchen design is efficiency. In most kitchens, activities center around (1) food storage (dry, refrigerated, and frozen), (2) food preparation (preparing, cooking, and serving), and (3) cleaning (cleaning, dish washing, and waste recycling). The area for each of these should be carefully considered in kitchen design in order to maintain continuity in the sequence of events that are related to the various activities. Each of these areas should be relatively contained. Each area should be well planned for the use of energy. Figure 4-5 illustrates a prototype kitchen design where areas of energy consumption can clearly be seen.

Locating the kitchen within the house is dependent upon many factors. Since the equipment, devices, and lighting are used quite often, there is a large amount of internal heat generated in the kitchen. This may mean that a central location for the kitchen is most desirable in houses in temperate or cool climates. Since the kitchen is used a great deal throughout

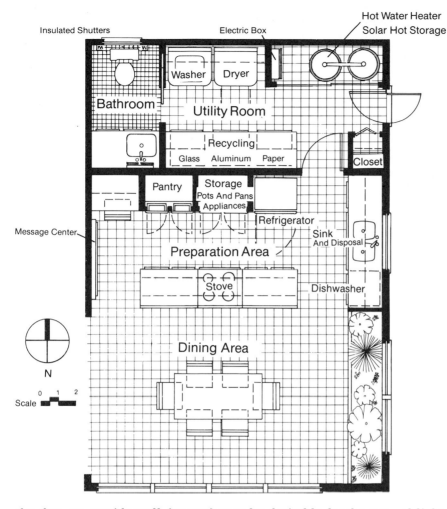

Insulated Shutters   Electric Box   Hot Water Heater
                                              Solar Hot Storage

Washer   Dryer

Bathroom   Utility Room

Recycling

Glass   Aluminum   Paper

Closet

Message Center

Pantry   Storage
          Pots And Pans
          Appliances

Refrigerator

Sink
And Disposal

Preparation Area

Stove

Dishwasher

N

Scale   0   1   2

Dining Area

FIGURE 4-5 Kitchen service center. (Student project, University of Colorado at Boulder, Lisa Price.)

the day, an outside wall (or two) may be desirable for the natural light and view to the yard or garden. If the kitchen generally is used more at one time of the day, perhaps a particular orientation may be more desirable. Often kitchens are located at the southeast corner of the home for early morning and daytime use. With pantries being more popular, there is another reason for location adjacent to an outside wall where natural cooling can be used.

The day-to-day function of food preparation along with all the tasks that are associated with it involves a large amount of resources. Many foods and products are bought, used, and made into waste. For this reason, recycling should be planned for the kitchen. One primary recycling center should be organized and designed there in order to make it more efficient and less time-consuming. Paper, plastic, glass, aluminum, and organic wastes can all be collected and recycled. In order for recycling to really be accomplished, it needs to be convenient and visible. Waste must be easy to move. Once again refer to the kitchen plan for the location of the recycling center. Typical of the kinds of wastes of a home are the following:

1. Organic food waste
2. Food-preparation waste
3. Cans, bottles, and other containers
4. Dust, ash, and sweepings
5. Dirty water (from washer, shower, sink, etc.)
6. Human waste: feces and urine
7. Large items: appliances, boxes, furniture, etc.
8. Yard waste: grass, leaves, dead plants, fallen branches

The utility room deserves some attention in the design of a house. Often it is designed too small and placed in an undesirable location. Families with young children use the utility room several times a day. Utility rooms should be designed to receive solar energy for the potential of natural clothes drying and space heating. Consideration should be given to all the activities associated with clothes washing. This includes washing, drying, ironing, folding, and storage. Utility rooms can be thermally zoned so that when they are not in use, they can have lower temperatures in the winter.

A look at each of the major activities of the home in relation to the time of day when they are done may give some insight as to how to organize areas of the home into compatible thermal zones. Each household may be different in this respect; however, some similarities may occur. For example, dividing the day into four 6-hour periods usually reveals different sets of activities and associated energy needs. Early morning is the time when most of us are sleeping. Energy needs are generally low. Late morning, from 6 a.m. to noon, is full of activity: waking, washing, eating, working, etc. Afternoon, depending upon the day of the week, is slightly lower in activity. And night, from 6 p.m. to midnight, is another period of relatively high activity with many energy demands: food preparation, washing, lighting, television, etc.

SHELTER DESIGN

There are many forms of the housing unit. It can be attached or detached; it can be large or small; and it can be either owned or rented. It can be rural or urban or anything in between. If you trace people's changing shelter needs from childhood to old age, you see a remarkable diversity, which is reflected in shelter design. Take, for example, the starter home of a young family in the United States. Usually the home is in a suburban location, it is small, and it has two or three bedrooms. Life is internally oriented. As time goes on, the family outgrows the small home and has need for a house with greater area, more privacy, and more outdoor space. Life is expanding. When the children reach working or college age, perhaps, they will live in a dormitory or a small apartment. When they move into professional life, they may find a mate and move into a larger apartment or even a small house. As the parents grow older, they are probably living alone in an apartment, condominium, town house, or even a large suburban home. With old age, their needs become more dependent upon cooperative environments. This example, surely, does not represent every family pattern, but it is useful in demonstrating the notion of multiple shelter needs.

The cycle continues as we move in and out of varying levels of exterior and interior space needs. The available housing choice is limited by the existing stock of houses or the cost and availability of land, appropriate zoning, and cost of construction, and interest rates for new housing. As a result, it is difficult to stereotype our shelter needs into a simple, all-encompassing model. There are various forms of housing in response to the varying needs. The range of housing forms is responsive to considerations of density, economics, and architecture. Presently there are several trends in shelter design aimed at satisfying the ongoing housing demand. Shelter *needs* and *forms* are the subjects of the following sections.

Shelter Needs

Before we focus on forms and current trends in housing, especially those influenced by energy and cost, we will examine needs associated with shelter design. This allows us to view the central housing issues without getting overly involved with style, regional character, marketing, sales, or aesthetics. According to Sam Davis, editor of *The Form of Housing*, "Although supplying shelter for everyone is critical, the quantity of dwellings does not in itself solve the housing problem. Dwellings must meet the expectations and living modes of the user."[18] There is no question that the modes of living are varied. They vary not only according to individual changes or cycles in life but from person to person, family to family, and culture to culture.

A simple way to organize shelter needs is into two categories: envelope needs and internal needs. *Envelope needs* grow out of the interactions between the inside and outside, especially where climate affects comfort. Both positive and negative interactions occur (from fresh air to freezing weather). The *internal needs* are related primarily to the inside activities, their relationships with one another, and their dependency upon the various support systems. An important issue, which has arisen because of more tightly built buildings, is indoor air pollution, which is discussed in this section. Both the envelope and the internal needs address spatial and functional concerns, and both relate to the user. Figure 4-6 illustrates, in diagram form, the nature of envelope needs.

The envelope, or "skin," of the house gives us protection from the exterior elements. In more urban situations, this includes protection from one another as well as from the effects of climate and weather. As can be seen in Figure 4-6, there are five areas of interaction as described by James Marston Fitch. Interactions are atmospheric, luminous, sonic, bio-

INTERNAL ENVIRONMENT   WALL   EXTERNAL ENVIRONMENT

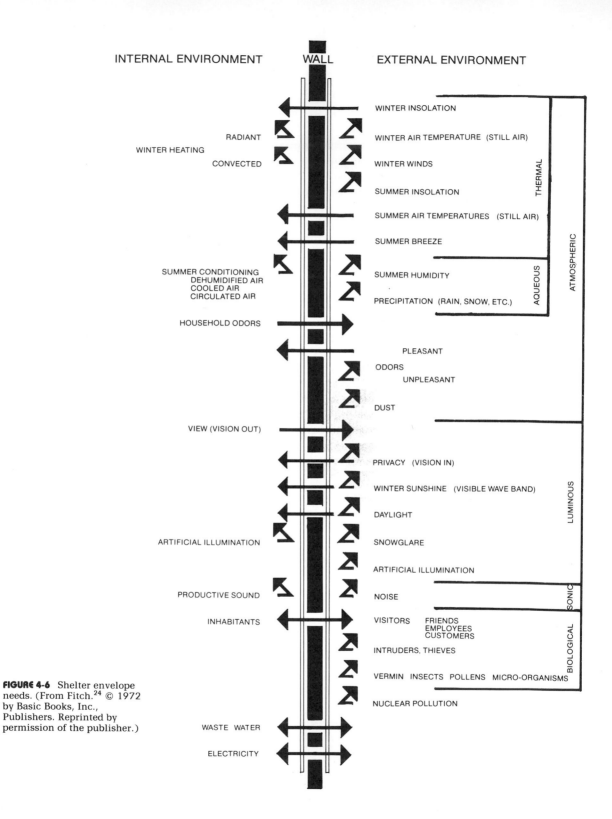

FIGURE 4-6 Shelter envelope needs. (From Fitch.[24] © 1972 by Basic Books, Inc., Publishers. Reprinted by permission of the publisher.)

logical, and social. The wall or envelope is not a static or impermeable object but a sophisticated filter through which the interior and exterior environments interact. From the first two areas come our greatest energy needs—the needs for heat and light. In this respect, shelter as passive solar collector has a great deal of appeal. Many new materials, methods of construction, and devices have recently added to the possibilities of improved envelope design. Providing greater resistance to skin heat loss and enhancing the greenhouse effect in concert make a more efficient passive system. Sun tempering, movable insulations, the heat mirror, and tighter construction are all adding to improved envelope design. Windows are again being considered as important architectural elements of the skin. According to Donald Watson, AIA, a window performs many functions and there are many factors that influence its design—from natural lighting, ventilation, and thermal control to security, noise control, and "regrettably for the designer of sunscreens in some areas," pigeons.

Ideally the skin is going to provide protection and resistance to the external elements or it is going to regulate them. Greater resistance comes with increased insulation and infiltration control. Additional roof, wall, and foundation insulation can easily be accomplished with new construction. With existing buildings it is more difficult and may not be cost-effective. Infiltration can be reduced in several ways according to the following percentages given by the American Society of Heating, Refrigerating, and Air-Conditioning Engineers (ASHRAE). Batt insulation placed around the perimeter between the foundation wall and the sole plate can reduce infiltration by as much as 25 percent. Packing wall outlets with insulation can reduce it by 20 percent. Controlling infiltration in ducts and vents can reduce it by around 20 percent. Although there is relatively little infiltration around the windows (this really depends on the type and age of the window), about 12 percent, movable insulation can be cost-effective when related to conduction and radiation losses and gains. Other infiltration losses include door losses (5 percent), fireplace losses (5 percent), and miscellaneous losses (around 13 percent).

Depending upon the climate, we spend as much as 80 percent of our time indoors. A great deal of this time is spent in the home. Many factors affect our experience there—the basic activities and internal functions of the home, the desired relationships among these activities, the ways in which the activities interface with the outside, the indoor furniture and props, and color and light, to name a few. Energy is important in several ways. Adequate passive solar space heating along with active hot-water heating should generally be employed. This requires proper orientation of interior thermal zones and zone coupling as discussed in Chapter 2. Thermal storage is important to the internal function and comfort at night. With the use of more and more electronic devices comes the need for sound and noise insulation. The need for higher internal air quality is beginning to be a problem with the tighter, more energy-efficient homes.

While the home has evolved to provide a more sophisticated response

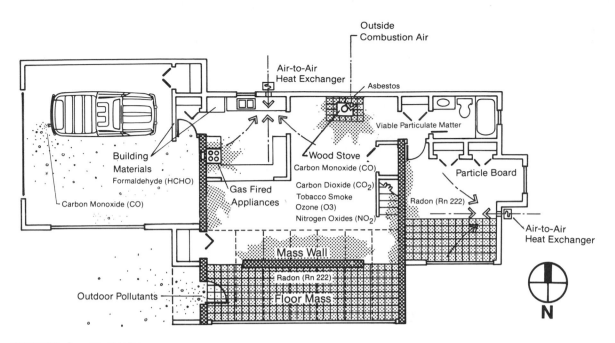

Outside Combustion Air

Air-to-Air Heat Exchanger

Asbestos

Viable Particulate Matter

Building Materials

Formaldehyde (HCHO)

Carbon Monoxide (CO)

Gas Fired Appliances

Wood Stove
Carbon Monoxide (CO)
Carbon Dioxide (CO$_2$)
Tobacco Smoke
Ozone (O3)
Nitrogen Oxides (NO$_2$)

Particle Board

Radon (Rn 222)

Air-to-Air Heat Exchanger

Mass Wall

Radon (Rn 222)

Floor Mass

Outdoor Pollutants

N

FIGURE 4-7 Sources of indoor air pollution. The floor plan indicates sources of indoor air pollution. The kitchen, mechanical room, and fireplaces or wood-burning stoves are prime areas of higher concentrations of air pollutants.

to the *outside*, we are, in fact, creating some new problems on the *inside*. Increased incidence of nausea, headache, disturbed sleep, irritation to the eyes, respiratory illness, bronchitis, heart disease, cancer, and even death are some of the health problems caused by certain construction materials in tightly built homes. In older homes with higher infiltration rates, many of the gases produced by various materials are displaced, in reasonable periods of time, with clean air and therefore present no real threat. With the newer, more energy-efficient homes, infiltration is reduced, and over time the gases increase in concentration.

Typical of the kinds of chemicals and gases that have been found to be harmful are carbon monoxide (CO), nitrogen oxides (NO$_x$), tobacco smoke, asbestos, formaldehyde (HCHO), ozone (O$_3$), radon (^{222}Rn), increased amounts of carbon dioxide (CO$_2$), and viable particulate matter, or total viable particles (TVP). According to Wadden and Scheff, coauthors of *Indoor Air Pollution,* "The quality of air we breathe and the attendant consequences for human health are influenced by a variety of factors. These include hazardous material discharges indoors and outdoors, meteorological and ventilation conditions, and pollutant decay and removal process."[73] The following is a list of major sources of indoor air pollution:

1. Gas-fired appliances (range, oven, pilot lights)
2. Wood stoves and fireplaces
3. Building materials (rock, sand, and clay products)
4. Particle board
5. Urea-formaldehyde foam
6. Asbestos fiber (spray or boards)

7. Human beings (smokers)
8. Outdoor pollutants (air pollution)

Figure 4-7 illustrates in floor plan the probable locations of these pollutants. It also suggests locations for air-to-air heat exchangers.

The awareness that many construction materials are unhealthy is relatively new, so the solutions at this stage of the process may be incomplete. However, there are several ways in which indoor air pollution can be reduced. The first way, and probably the most obvious way, is not to use materials and combustion processes that produce contaminants at harmful concentrations. This may not always be possible. In energy-conserving homes, for example, thermal storage in the form of masonry, concrete, or rock beds is desirable. The second way is to use ventilation rates and air-circulation systems to provide for adequate levels of oxygen and the removal of contaminated air. The third way is to use air filters and cleansers for recirculated inside air when outside air is either contaminated or too cold. And the fourth way, in very tight homes in cold climates, is to use air-to-air heat exchangers. Figure 4-8 is a diagram of an air-to-air heat exchanger.

Shelter Forms

Dwelling units vary greatly and can range from large single-family detached houses to small high-rise apartments. Differences in dwelling-unit types include unit size, unit ownership, unit location, relationship to the ground, adjacency to other dwelling units, number of outside walls, number of floors, relationship to parking or garage space, and types of energy systems. The list that follows describes a range of eight dwelling-unit types. They include the single-family detached unit, single-family zero-lot-line unit, duplex, row house, quadruplex, three-story walk-up, slab-block apartments, and high-rise apartments.

1. **Single-Family Detached Unit.** This is the most common dwelling type in the United States. Usually it is located in a low-density zone; is one-story, two-story, or bi-level, has four or more exposed exterior walls; and is sited on grade. The single-family lot size is 5000 square feet or larger; the density can reach as high as 4 to 8 units per acre; and the unit size is typically 1700 square feet or larger. Individual energy systems are used.

2. **Zero-Lot-Line Unit.** This is similar to the single-family detached dwelling except that the zero-lot-line unit is located along a property line. Typically one or two stories in height, it has three or four exposed exterior walls. The lot size is around 5000 or 7000 square feet. There are usually 6 to 10 units per acre. And the unit size is 1200 to 1800 square feet. Individual energy systems are used.

3. **Duplex.** Similar to the zero-lot-line house, the duplex has a shared or party wall with an adjacent dwelling unit. Typically one or two stories in height, the duplex has three exposed exterior walls. The lot size is 4000 or 5000 square feet. The density usually is between 6 and 10

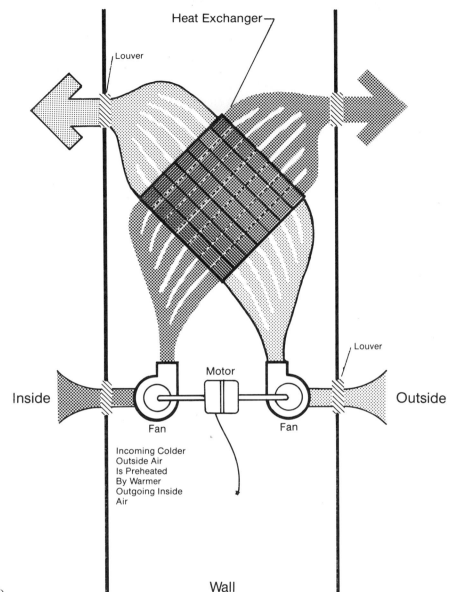

Heat Exchanger

Louver

Louver

Inside

Outside

Motor

Fan

Fan

Incoming Colder
Outside Air
Is Preheated
By Warmer
Outgoing Inside
Air

FIGURE 4-8 Air-to-air heat exchanger. (From Shurcliff.[52])

Wall

units per acre, and the unit size is 1200 to 1800 square feet. Individual or shared energy systems are used.

4. **Row House.** The row house or town house has units placed adjacent to one another with two shared or party walls and two walls and a roof exposed to the exterior. The density ranges between 12 and 18 units per acre. The lot size is between 2000 and 3000 square feet. Typically, the row house is two or three stories high with access on grade. Unit sizes are between 1200 and 1500 square feet. Individual energy systems are used.

5. **Quadruplex.** The quadruplex, like the duplex, has four attached units ranging in size between 1200 and 1500 square feet. The density ranges between 16 and 24 units per acre with lot sizes approximately 6000 square feet. Units share walls or floors and ceilings. Energy systems are either individual or shared.
6. **Three-Story Walk-Up.** Stacked apartments or individually owned units with stair access only (no elevator). The density ranges between 25 and 40 units per acre with unit sizes approximately 800 to 1000 square feet. Energy systems are typically centralized and individually metered.
7. **Slab-Block Apartments.** Stacked apartment units with a majority of the units above the ground (four to eight stories). The density is between 50 and 75 units per acre. Unit sizes are approximately 800 to 1000 square feet. This dwelling type requires elevators. Energy systems are usually centralized and individually metered.
8. **High-Rise Apartments.** Stacked units of approximately 800 square feet. Typically all the units are above the ground. Densities range between 50 and 120 units per acre. This dwelling type requires elevators. Energy systems are centralized.

Despite the variety of possible housing-unit types, in the United States the single-family detached unit with its own piece of land remains the most popular housing form. As a result, 70 percent of the dwellings in the United States are single-family houses.

The popularity of the single-family detached shelter design is fostered by several features that are inherent to its form. This form offers the greatest potential for individual or family identity. Individual and family philosophies, values, and interests can be expressed and supported by the single-family house. This form provides convenient access to the automobile (often more than one automobile). Both attached and detached garages, carports, and parking areas are easily accommodated. This form promotes privacy and security. Privately owned space surrounding the home forms a protective barrier for the intimate activities of home life. This form supports child supervision both indoors and outdoors. A fenced-in yard in view of the kitchen is extremely necessary for families with young children. Although the single-family house is generally more energy-intensive, people are often eager to incorporate energy conservation. This is due to the fact that most single-family homes are owned by the occupants, who are therefore more inclined to make the physical changes so they can enjoy the reduced energy costs.

The single-family house is not without its problems. First, it necessitates sparse development. This kind of development is area-intensive, requiring large areas of land for housing plots, roads, utility easements, and open space. Second, the cost of this form of housing is high and growing even higher. In fact, today the cost of new single-family housing is out of reach for most families, with high interest rates and high energy, land, and construction costs. Third, single-family development requires more materials, natural resources, and services. For example, the water needs of a family living in a single-family dwelling are substantially greater than the needs of a family living in a high rise. Community-service needs are greater—snow removal, garbage collection, fire protection, utility distribution lines, etc.

Moving to other forms of housing can take away some of the benefits of single-family housing. Usually there is a reduction in private outdoor space; there is usually a loss in unit identity; there is reduced privacy and security; and there is less convenience to the automobile. Many developers have attempted to offset these losses with certain amenities: larger communal open spaces with recreational facilities, less yard or garden maintenance for those who do not desire the responsibility, and a closer relationship to the city or to community activities and services. These other forms of housing vary according to the numbers of units and their relationship to each other and the ground. The percentage of housing starts for dwelling types other than single-family is increasing. Multifamily housing will be discussed in more detail in the next two chapters.

Single-family development, which represents a little over 50 percent of the present housing market, is changing. The trends are to bring overall costs down and to respond to contemporary needs. Many demonstration programs and national competitions have been conducted, such as the NAHB and Solar Energy Research Institute (SERI) demonstration program, the U.S. Department of Housing and Urban Development design awards program, and the competition sponsored by the American Plywood Association, *Progressive Architecture,* and *Better Homes & Gardens.* To summarize the trends, single-family houses are becoming smaller, are being located on smaller land plots, and are being made more energy-efficient. A more detailed list follows:

1. Smaller land plots (better use of outside space)
2. Smaller floor area (with the average around 1700 square feet)
3. Less waste of space and materials
4. More free-flowing traffic patterns
5. Good functional interior zoning (with open planning)
6. More energy conservation (more insulation, better infiltration control, etc.)
7. Use of solar energy systems (passive space heating and active domestic hot-water heating)
8. Sun tempering (more glazing on the south and less on the north)
9. The use of earth sheltering
10. Designing for lower maintenance
11. Contrasting ceiling heights
12. The use of more efficient operable windows for better insulation and ventilation
13. The use of more natural materials inside and outside (especially low-energy-intensive ones)
14. The greater use of mass production techniques and materials

These trends are supported by several demographic transformations.

First, there is a rising older population. According to the Urban Land Institute, the average age in the year 2030 in the United States will be 37 years of age. This is considerably higher than it is now at 29 years of age. Second, there is a drop in the average household size while at the same time there is an increase in the number of new households. Presently there are less than three persons per household, and the number of households is increasing at a very high rate. Third, there is an increase in the number of women in the work force, which is associated with smaller household size. And fourth, there are migration patterns to the west and south and to smaller cities and rural areas. This is a reversal from trends of the recent past. Generally, these trends indicate that there will be continued development of the detached single-family house at least in the near future. As the average age of the population increases and as the value of money decreases, this form of housing is likely to change—to more livable attached units.

SINGLE-FAMILY HOUSING DEVELOPMENT

As long as detached single-family dwellings remain a major thrust in the near future of American housing, they should be examined and improved upon. A look at the land plot and dwelling, in light of their attractions and inherent problems and the current trends in design, should reveal changing patterns of this form of shelter design. In relation to energy, many planning and design changes can reduce consumption and reliance upon natural resources. From site selection and subdivision to land use and landscaping, the small-scale development can be more self-sufficient, productive, and attractive. Two development types or scales that are associated with detached single-family housing are described here— *single-lot development* and *small-parcel development.*

Single-Lot Development

Although single-lot development is decreasing because of the high costs of property and building construction, it is important because, to date, most of the energy-responsive design has occurred with this form. It is projected that there may be 200,000 solar housing starts by 1985, and many will occur as single-lot developments. Unique to this form of development, is the specific relationship of design to the land and lifestyle of the owner. As a result, land use, landscape design, and building form can be extremely sensitive to issues of quality of life. This includes energy. People are interested in saving money from reduced energy consumption and enjoying the benefits of natural light and heat, garden and greenhouse environments, hot tubs, etc.

Land use for single-lot development is not complicated. It involves a rather basic organization of functions. Single-lot sizes vary from about a sixth of an acre (7000 square feet) to as much as 35 acres. With zero-lot-line design, even smaller lots can be developed. With the smaller lots, the main house is located near the center of the lot, within a buildable

FIGURE 4-9 Summer photograph. (Courtesy of Phillip Tabb Architects.)

area, usually creating front and rear yards. The garage is either attached or detached. The front yard is used for formal entry and a transition to public space, and the rear yard is reserved for more informal outdoor activities. Adjacent or attached to the house are patios, decks, balconies, or courts that act as extensions from the house envelope into the surrounding environment. Typical of the activities that occur in the yard are sunning, outdoor eating, gardening and landscaping, and adult and child playing and swimming.

Figures 4-9, 4-10, and 4-11 illustrate a single-lot design. The photographs are taken in summer and winter, illustrating the vast difference from season to season in a temperate zone. Note the difference in branch

FIGURE 4-10 Winter
photograph. (Courtesy of
Phillip Tabb Architects.)

density of the trees, even though in the winter photograph there is a lot
of snow on the bare branches. The lot measures 70 feet by 120 feet. The
garage located to the south of the lot has been converted to a studio. A
sunspace and deck are located on the south of the main house. Stepping
planter beds join the garden with the basement through the sunspace.
The utility yard is to the east, for clothes drying, compost heaps, and outdoor storage. The garden is to the west, with shade trees, fruit trees,
shrubs, and bushes.

Energy-oriented landscaping, including earth sheltering, is becoming
more popular. It represents an incremental approach to energy conservation that can be implemented over any period of time. The single-lot

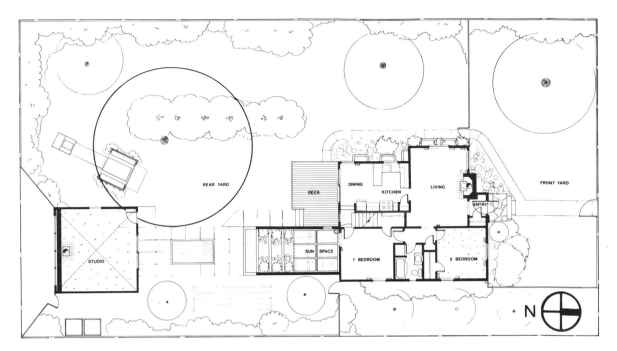

Labels within the figure: REAR YARD, DECK, DINING, KITCHEN, LIVING, FRONT YARD, ENTRY, STUDIO, SUN SPACE, 1 BEDROOM, 2 BEDROOM, N

FIGURE 4-11 Single-plot site plan. (Courtesy of Phillip Tabb Architects.)

development has plenty of open space in which to create a variety of landscape designs using both native and introduced species. Both overheating and underheating can be augmented with proper landscaping. According to Anne Moffat and Marc Schiler, authors of *Landscape Design That Saves Energy*, ''Landscaping to save energy is a powerful tool for conservation: it controls wind, solar radiation, and precipitation, tempers extremes of climate, and can save up to 30 percent of a home's total energy requirements for space heating and cooling.''[43] This is quite significant when coupled to the many solar energy and conservation techniques that can be incorporated with the house itself. Energy savings, of course, depend upon the severity of the climate, the local cost of conventional fuels, and the quality of the landscape designs. The following is a listing of general energy-oriented landscaping techniques. Application of these techniques should be carefully considered for specific climates and locations.

1. **Summer Shading.** Well-placed and -selected species of trees, bushes, and shrubs near south and west facades and on the roof can reduce the need for air-conditioning. Landscaping elements can be positioned to funnel breezes for additional cooling.
2. **Windbreaks.** Properly designed windbreaks can reduce both conduction and infiltration losses to the envelope as heat loss at the surface is proportional to the square of the wind's velocity. Shrubs and vines on a building surface can help create still air pockets.
3. **South Surface.** The ground surface to the south of a house can be used to increase solar collection through reflection of solar energy to

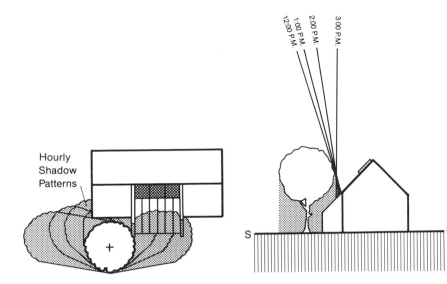

12:00 P.M.
1:00 P.M.
2:00 P.M.
3:00 P.M.

Hourly
Shadow
Patterns

S   N

FIGURE 4-12 (Left) Landscape shading plan.

FIGURE 4-13 (Right) Landscape shading section.

windows or collectors. As much as 20 to 30 percent of additional solar energy can be gained. In summer, ground cover can reduce the amount of reflection and building heat gain.

4. **Earth Sheltering.** Earth sheltering the windward, northern, and roof portions of a house can reduce conduction and infiltration and create more stable temperature swings. Earth sheltering is especially attractive with sloped sites. Earth berming can also be used to help buffer prevailing winter winds before they reach the house.

5. **Food Production.** On-site food production can easily be accomplished. The type and quantity of food are dependent upon the length of the growing season, amount of sunlight, availability of water, quality of the soil, and pest control. Many common vegetables can typically be produced in sufficient quantity to reduce dependence upon outside sources. Solar greenhouses can also support a certain amount of food production on a year-round basis.

6. **Water.** The use of water in landscaping is underestimated. Natural evaporative cooling is appropriate in hot-arid and temperate climates. Water not only reduces overheating but adds humidity to the air. It is also an extremely attractive landscaping element. Water is being introduced indoors in the form of fountains, hot tubs, and spas.

Summer shading is a popular landscaping technique. The proper location of trees can not only offset intense solar radiation during the summer, it can provide a pleasant quality of light and sense of place. Locating trees for shading is a slightly different site-landscaping process than ordinary landscaping. Often the placement of deciduous trees for summer shading needs to be closer to the south side of the house than intuition seems to indicate. Figures 4-12 and 4-13 illustrate in plan and section the proper location of shade trees and indicate the summer and winter

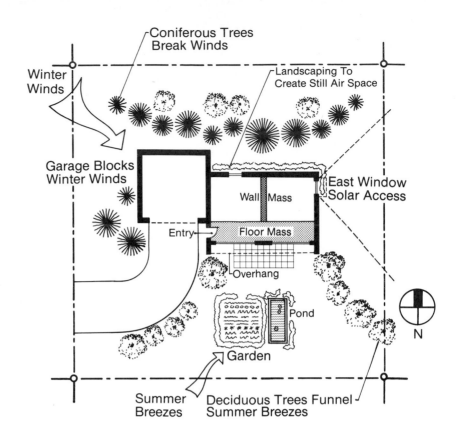

Coniferous Trees
Break Winds

Winter
Winds

Landscaping To
Create Still Air Space

Garage Blocks
Winter Winds

Wall Mass

East Window
Solar Access

Entry

Floor Mass

Overhang

Pond

Garden

N

Summer
Breezes

Deciduous Trees Funnel
Summer Breezes

FIGURE 4-14 Temperate-zone site plan.

shadows. Care should be taken in selecting the species of shade tree, as the various trees have different branch densities (an important consideration if winter solar radiation is needed) and leaf densities (for summer shading).

In the continental United States, the climatic conditions, like the geographic conditions, are varied and contrasting—the wet Olympic peninsula, the California coast, the hot-arid deserts of the southwest, the colder mountain ranges, the semiarid Great Plains, the Mississippi valley, the northern areas around the Great Lakes and New England, the Atlantic coast, and the hot-humid areas around the gulf coast. Although there are, in fact, many different climates, most climate-adapted design scenarios fall into four zones: temperate, cool, hot-arid, and hot-humid. Detailed descriptions of these climate zones can be found in *Design With Climate*, by Victor Olgyay.[48] In using the design techniques suggested for these four zones, care should be taken when applying them to specific locations and sites. Simplified shelter designs and energy-oriented landscape concepts are here discussed and illustrated for each of these four climate zones.

The _temperate zone_ is characterized by contrasting weather, with both overheating and underheating. Temperature extremes are not as great as those experienced in cold or hot regions. However, both diurnal and sea-

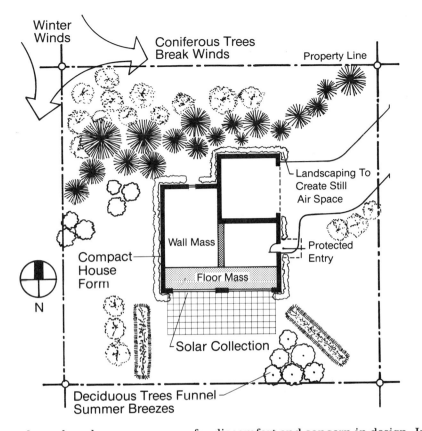

Winter Winds

Coniferous Trees Break Winds

Property Line

Landscaping To Create Still Air Space

Wall Mass

Protected Entry

Compact House Form

Floor Mass

N

Solar Collection

Deciduous Trees Funnel Summer Breezes

FIGURE 4-15 Cool-zone site plan.

sonal weather changes are cause for discomfort and concern in design. In winter, energy conservation and solar collection are desirable; and in summer, conservation and shading are desirable. In the more arid temperate regions, evaporative cooling can be employed. Figure 4-14 illustrates a generic site plan for a single-lot development. The house shape is aspected generally along the east-west axis. Entry is to the south or east. The garage form protects the entry from prevailing winds as does a conifer tree windbreak to the northwest. Thermal floor mass is located along the south with quite a number of direct-gain windows. Insulated glass or shutters are used on the north and west windows. Overhangs are located over windows to the south for summer shading. Additionally, deciduous trees are positioned to the south, southwest, and west for summer shading.

The *cool zone* is characterized by cold weather, with a great deal of underheating during autumn, winter, and spring. During the long winter nights it is especially cold. Houses should be designed for solar gain, energy conservation, and protection from prevailing winds. Design techniques, in response to these weather conditions, should be more exaggerated for better performance. Figure 4-15 illustrates a generic site plan for the cool climate. The house aspect is square; in fact, the form is more optimal as a cube with minimal envelope heat loss. The entry should be

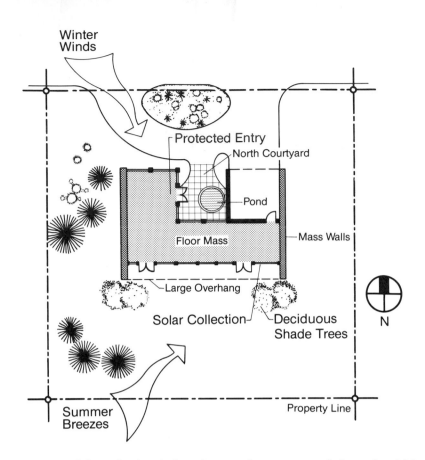

Winter
Winds

Protected Entry

North Courtyard

Pond

Floor Mass

Mass Walls

Large Overhang

Solar Collection

Deciduous
Shade Trees

N

Summer
Breezes

Property Line

FIGURE 4-16 Hot-arid-zone
site plan.

well protected from both wind and snow. Large areas of glass should be
used for solar gain. Thermal mass should be both distributed and concentrated, and it should be designed for longer periods of night. Night
insulation should be used to prevent night heat losses. Conifers should
be used for windbreaks. Greater attention should be paid to thermal zoning, temperature setbacks, and infiltration control. Earth sheltering is
very appropriate especially on south-sloping sites.

The _hot-arid zone_ is characterized by warm, dry weather. Because of
reduced cloud cover and moisture in the air, solar radiation gained during
the day is quickly reradiated at night (night radiation) back into the
atmosphere, thereby causing cold temperatures. Extreme temperature differences can be experienced diurnally. Figure 4-16 illustrates a generic
site plan for the hot-arid climate zone. In many hot-arid areas, winter daytime solar collection is needed to heat the house for the day and, especially, the night. For this reason, night thermal storage, associated with
solar heating, is desirable. Summer overheating is by far the most dominant weather condition. The house aspect should be slightly elongated
along the east-west axis and more open to the north. A U-shaped plan
with an open court to the north is appropriate. If shading can be accomplished with landscaping, it should cover the roof and the west facades.

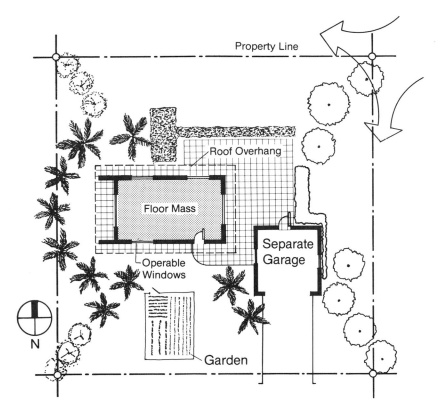

Property Line

Roof Overhang

Floor Mass

Operable
Windows

Separate
Garage

N

Garden

FIGURE 4-17 Hot-humid-zone
site plan.

Capturing natural breezes and incorporating evaporative cooling are also desirable. The key to climate-responsive design in this region is overcoming extremes during summer midday and winter night.

The *hot-humid zone* is characterized by warm, humid weather. Overheating, during both day and night, is typical of this zone. Stagnant, humid warm air dominates the weather. Therefore, means of encouraging and capturing breezes and reducing direct solar radiation are important. Figure 4-17 illustrates a generic site plan for this zone. The plan aspect should be elongated perpendicular to prevailing summer breezes. The garage or carport can be detached to help funnel breezes to the main house. The plan should be more complex and open, with windows equally placed around the envelope. Summer shading with shading devices and trees is desirable. Thermal mass should be distributed to prevent indoor air from overheating. The south-lot ground should be covered with ground material suitable for absorption of the solar radiation.

Table 4-2 suggests general objectives and adaptations for energy-responsive site planning for both individual buildings and small-parcel developments. It is based upon the broad climatic effects that are associated with the climate regions discussed—cool, temperate, hot-humid, and hot-arid. These suggestions should not be taken too literally, but they should serve as a guide or set of objectives during the initial site-planning design process.

TABLE 4-2
Climate-Oriented Site Planning[67]

	Cool	Temperate	Hot-Humid	Hot-Arid
		Objectives		
	Maximize warming effects of solar radiation. Reduce impact of winter wind. Avoid local climatic cold pockets	Maximize warming effects of sun in winter. Maximize shade in summer. Reduce impact of winter wind but allow air circulation in summer	Maximize shade. Maximize wind	Maximize shade late morning and all afternoon. Maximize humidity. Maximize air movement in summer
		Adaptations		
Position on slope	Low for wind shelter	Middle-upper for solar radiation exposure	High for wind	Low for cool air flow
Orientation on slope	South to southeast	South to southeast	South	East-southeast for p.m. shade
Relation to water	Near large body of water	Close to water, but avoid coastal fog	Near any water	On lee side of water
Preferred winds	Sheltered from north and west	Avoid continental cold winds	Sheltered from north	Exposed to prevailing winds
Clustering	Around sun pockets	Around a common, sunny terrace	Open to wind	Along E-W axis, for shade and wind
Building orientation*	Southeast	South to southeast	South, toward prevailing wind	South
Tree forms	Deciduous trees near building. Evergreens for windbreaks	Deciduous trees nearby on west. No evergreens near on south	High canopy trees. Deciduous trees near building	Trees overhanging roof if possible
Road orientation	Crosswise to winter wind	Crosswise to winter wind	Broad channel, E-W axis	Narrow, E-W axis
Materials coloration	Medium to dark	Medium	Light, especially for roof	Light on exposed surfaces, dark to avoid reflection

*Must be evaluated in terms of impact on solar collector size, efficiency, and tilt.

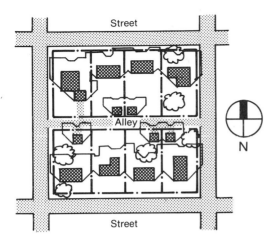

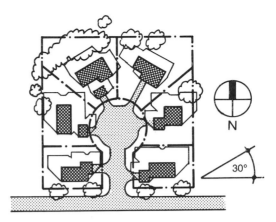

FIGURE 4-18 (Left) Small-parcel development, alternative 1: standard subdivision.

FIGURE 4-19 (Right) Small-parcel development, alternative 2: cul-de-sac.

Small-Parcel Development

A small parcel is differentiated from a single lot by the number of houses or lots being developed. A small parcel is typically subdivided into two to ten houses under one plat or construction contract. Shelter designs can be customized, but more often they are repetitive with built-in variations. To a certain extent, this is like the automobile industry, where basic models are introduced with many options to the base design. Specific elements of the design are open to customizing. In solar small-parcel development, these options may include domestic hot-water heating systems, solar greenhouses, night insulation, wood-burning stoves, etc. Energy conservation should be a part of the basic design. There are many freedoms and flexibilities with this form of development. Three subdivision alternatives follow and are explained and illustrated in Figures 4-18 through 4-20. They demonstrate different lot configurations, organizations, automobile accesses, and shared spaces.

The first example represents a standard subdivision that may fall within the Jeffersonian grid. The land parcel is bounded by a street and alley or by two streets. Usually the lots are long and narrow with fairly large front and rear yards and small side yards. The houses are structured more formally on the street side and more informally on the alley side. With this form of development, there is a clear separation between the front door and back door with attendant activities.

The second example is organized by a cul-de-sac with radiating lots. In this example, zero-lot-line development can reduce lot sizes. Automobile access and visitor parking are shared in the center of the development, while at the extremities in the rear of the lots, more private activities occur. In some instances, orientation to the south for optimum solar collection is not possible because of the radiating lot positions, but the orientation should be no more than 30° off-axis, as shown in Figure 4-19. This form of organization will be discussed in more detail in the next chapter.

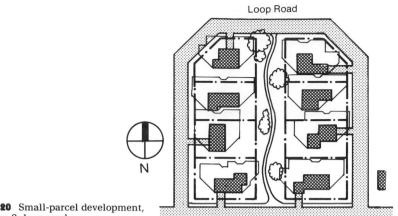

Loop Road

N

FIGURE 4-20 Small-parcel development, alternative 3: loop road.

The third example demonstrates a loop road, which is planned around the perimeter of the development. As a result, an internal open space that is adjacent to or shared by each parcel can be created. Here the center can be reserved for pedestrian activities or bike paths while automobile traffic is limited to the perimeter of the development. This form of development lends itself to curvilinear patterns of lot placement. This is the form used throughout the solar village at Davis, California.

Much has been written on the pros and cons of subdivision development. It represents the major development pattern for large numbers of single-family detached residences. Subdivisions can vary drastically in size from only a few families to populations of over 60,000, as in the case of Levittown. The form of subdivision design is generally a network of linear repetitions of street, sidewalk, utilities, plot (or two plots with a double-loaded arrangement), and, in some instances, open space. Subdivision design over the last several decades has broken away from the more rigid grid to curving rows of detached housing. This was done in part to help slow down automobile traffic. Nevertheless, to allow for adequate solar access and proper orientation to the sun, this pattern may have to be more restrained. The curves and undulating arrangement of houses can still occur, but they should be designed within the reasonable limits of solar collection.

SUMMARY

The primary purpose of residential settlement is to provide a supportive environment for the housing of people. Shelter design is the common denominator and basic element. As a facet of residential development, it offers the greatest potential for technological change. Solar technologies and energy-conservation techniques are bound to evolve to greater levels of efficiency, and shelter design is bound to change at a phenomenal rate. The concept of responsive shelter design is important no matter what the form of the housing unit. The facet of solar residential development offering the greatest challenge is multifamily housing, including cluster development. This land-planning concept clearly demonstrates an opportunity for saving costs, land, and energy and suggests new levels of cooperation and sharing.

The single-family house located on a large land parcel will remain the preferred choice, and subdivision will remain the vehicle. Yet the single-family house type will undergo terrific economic scrutiny as land-use strategies will move toward higher densities, mixed uses, and more homogeneous satellite communities. The evolving concerns at the shelter-design level need to impact the larger planning scales and vice versa. This process can be seen more clearly in the next chapter, "Cluster Development."

CHAPTER 5

CLUSTER DEVELOPMENT

Our wealth is inherently common and our common wealth can only increase, and it is increasing at a constantly self-accelerating rate.   R. BUCKMINSTER FULLER[27]

The disadvantages of single-family detached housing are causing planners and developers to look for alternative forms to meet the present housing shortage. We currently are short of adequate housing in the United States at a rate of 2 million units per year. The solutions must be affordable, and they must respond to the many complex needs and types of users. With this in mind, the cluster form is an attractive alternative. Cluster housing has certain economic advantages while it maintains many of the qualities of single-family detached housing. It is relatively low in density and can be very residential in character. The cluster form has inherent amenities while it preserves a reasonable level of privacy. With careful modifications to current cluster designs, this form should prove to be versatile and responsive to contemporary needs, including energy needs.

This chapter is the second in the sequence of discussion of residential planning scales. The cluster form of shelter design has certain potentials beyond those that are common to the single housing unit. The cluster, with multiple housing units, has many new problems and challenges. It is the purpose of this chapter to examine cluster development in relation to its ability to provide shelter and conserve energy. This includes the unique needs and form responses of *cluster development*, its ability to save energy through *shared energy systems*, and the patterns of growth or the methods of *cloning clusters*.

FIGURE 5-1 Social housing cluster, Chambéry, France. This project is the result of a French national competition for low-cost solar housing. (Courtesy of Palloix, Patriarche & Chavin.)

(a)

CLUSTER DEVELOPMENT

The rationale behind cluster development is simple. It is a compromise between the economic advantages of higher-density housing and the personal values held by single users or families of lower-density housing. A cluster can vary in size and density. However, the cluster most suited to blend the detached housing form with increased density is the low-rise, medium-density cluster. At an average density of 15 dwelling units per acre, this type of cluster development is over 3 times as dense as the conventional single-family detached housing development, which has an average density of 4.5 dwelling units per acre. The low-rise, medium-density cluster is usually a maximum of four stories in height and, therefore, can easily retain the residential scale and character of the detached single-family development.

Figure 5-1 illustrates a government-subsidized cluster housing development in Chambéry which is located in eastern France. This project is a good example of the low-rise, medium-density cluster concept. Often with low-rise cluster development, such as this, the massing of housing units is less regimented, resulting in a more individual identity and char-

(b)

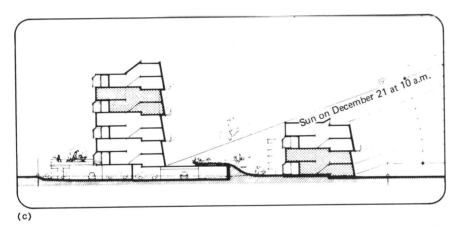

Sun on December 21 at 10 a.m.

(c)

acter. Each of the fifty-three housing units in this development has its own integral active and passive solar collection system. Most of the dwellings have two stories, allowing for spatial variety. The active collector arrays are angled toward the winter sun, while the passive collectors are vertical windows and glass doors. On top of the parking garage is an outdoor communal space. A result of a national competition in France, this project is recognized for its simplicity and low construction cost.

The form of the cluster has two major effects upon its ability to utilize solar energy. The first effect, which is a positive one, is that cluster housing is generally more energy-conserving. This is due to the closer packing of housing units with many shared walls. As a consequence, there is a reduction in the heat load of each unit. This should reduce the area requirements of an active or passive system. The second effect is that in cluster housing, which is usually center-oriented (that is, each housing unit is oriented to a central place or courtyard), solar-orientation or solar-access inequities can occur. Based upon the limitations of solar energy, it is difficult to design adequate solar heating systems for all of the units

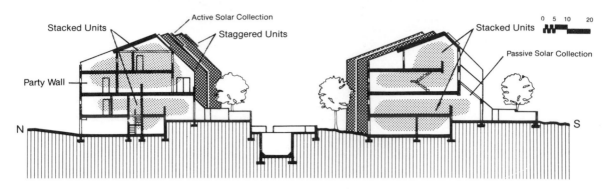

Stacked Units

Active Solar Collection

Staggered Units

Stacked Units

0 5 10 20

Passive Solar Collection

Party Wall

N

S

FIGURE 5-2 Economies of cluster housing.

in a circular configuration. Many of the units are positioned with orientations that are uneconomical for solar heating. These effects are examined in detail after discussion of cluster needs, concepts, and characteristics.

Cluster-Development Needs

Cluster development can be made more viable with greater emphasis and response to certain human needs. These needs vary and often are in contrast to one another—for example, the need for privacy versus the need for community. As we learn more about human behavior and the needs associated with shelter, neighborhood, and community, more articulated design responses can be made. The hierarchy and priority of needs will vary from time to time. This necessitates careful analysis of contemporary needs and the economics involved in responding to these needs. This should lead to more sensitive design decisions. Presently there are several shelter needs critical to development of cluster housing. Four general categories of needs are apparent: *economy, privacy, convenience,* and *amenities.* Each category contains specific needs and many detailed considerations.

The economy of cluster development needs careful analysis. Low-rise cluster housing is very similar to single-family detached housing and, therefore, has many of the added costs. The economic values of simplicity, repetition, stacking, prefabrication, etc. are often difficult to achieve with cluster housing. Many hidden costs arise because of the need to create privacy, convenience, and amenities to attract the home buyers. Providing affordable housing is one of the prime reasons for the interest in this form. The economic benefits to cluster housing are (1) lower land costs, that is, lower land cost per dwelling unit; (2) lower construction costs, that is, lower labor and material cost per dwelling unit; and (3) lower operating costs per unit for maintenance, energy systems, and landscaping.

Further economies can be achieved by avoiding difficult sites where topography is too varied or adjacent properties have negative environmental influences. Avoid overly complicated building designs with complex footprints. Excessive corners or unconventional geometry can increase both labor and material costs. Take a serious look at unit sizes.

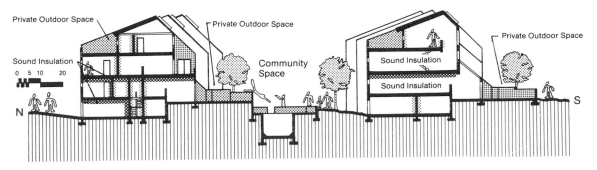

Private Outdoor Space

Private Outdoor Space

Private Outdoor Space

Sound Insulation

0 5 10   20

Community Space

Sound Insulation

Sound Insulation

N

S

FIGURE 5-3 Privacy in cluster housing.

Savings can come from smaller unit sizes; however, reduction should be based on careful analysis of internal and marketing needs. With rising conventional fuel costs, effective energy-conservation measures and solar heating systems can accrue savings. Typically, at least two sides of most housing units within a cluster are shared with minimal heat loss. The quality of the development can have a large impact on the economics of the project. Consider the quality of the context and the particular group of users for which the development is intended. Amenities that may initially attract buyers may be costly and difficult to maintain. Figure 5-2 illustrates some of the economies of cluster housing.

The need for privacy is probably the single most important issue facing livable cluster housing. With an increase in density, from 4.5 units per acre with detached single-family housing to 15 units per acre with cluster housing, the ability to maintain privacy is much more difficult. With cluster housing, individuals and families live in closer proximity to one another. A wall, floor, or ceiling separates one unit from another. The lack of adequate protection from sounds, vibrations, and odors and the lack of adequate visual, physical, and psychological privacy can contribute to increased stress and tension for the occupants of higher-density housing—something to avoid at nearly any cost. In cluster housing there are many shared indoor and outdoor activities that can contribute to further complications in securing privacy. Both traffic and noise must be treated as invaders that can interfere with privacy—the need for withdrawal, concentration, self-reliance, solitude, and meditation.

To alleviate privacy problems in cluster housing, several measures can be taken. Establish clear individual and family boundaries or territory. Ownership or occupancy bounds should be easily identified and reinforced with fences, walls, landscaping, changes of material, etc. At the internal level, provide the potential for functional separations of people from one another. Provide clear identification of entries—both front, or formal, entries and back, or informal, entries. Provide semiprivate sequences of spaces with protected entries. Provide adequate outdoor private space for each unit—provide as much visual privacy as possible. With housing units adjacent to public or semipublic areas, care should be taken with entry, door, and window placement and design. Shared barriers should be well-soundproofed, especially those that are indoors. Refer to Figure 5-3.

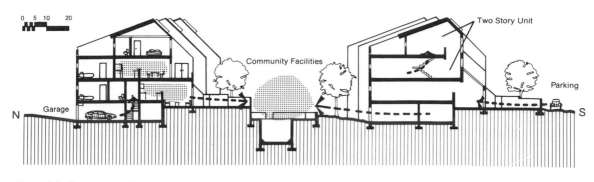

Figure 5-4 Convenience in cluster housing.

The need for convenience is probably not as critical to life support as some of the other stated needs; however, it is a carryover from the lifestyle associated with single-family detached housing. Convenience and accessibility can contribute to reducing stress. For this reason, it is an important cluster design determinant. The design and location of certain areas such as the kitchen, laundry, linen closet, and broom closet can have a large impact on the ability of the occupants to maintain food preparation and washing—two functions involving a large number of interrelated activities. Small inconveniences can be cause for irritation and discontent, especially if experienced day in and day out. (Refer to the discussion of kitchen design in Chapter 4.) Indoor spaces such as the kitchen, dining room, and living room should have a close relationship to private outdoor spaces that may be used for summer outdoor eating or children's play.

The relationship between the automobile and each individual housing unit requires careful consideration. In single-family detached housing, the automobile is stored in an attached or detached garage or carport or on a nearby street. In any case the car is usually located in close proximity to the house. This facilitates easy movement of groceries, furniture, garbage, etc. in and out of the house and a swift transition between the house and the car during inclement weather. Other activities, such as mail delivery, milk delivery, and garbage pickup, should be planned for the convenience of the occupants. Conflict usually occurs between the need for convenience and the need for separation of automobile access and parking with strict pedestrian areas within the cluster development. Figure 5-4 illustrates some of the issues of convenience and accessibility.

The provision of appropriate amenities can make cluster housing more livable. The quality and quantity of amenities are a practical function of the number of occupants of the cluster and their general income level. Generally the greater the population in a cluster development, the greater are the amenities and shared activities. Typically the kinds of amenities associated with cluster housing include swimming pools, tennis courts, and other provisions for outdoor recreation, saunas, hot tubs, and other indoor recreation facilities, day-care facilities, laundries, party rooms,

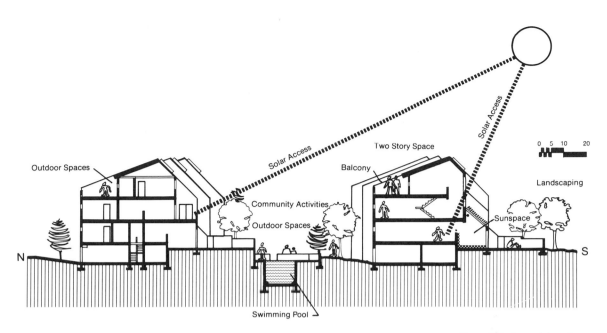

Figure 5-5 Amenities of cluster housing.

children's playgrounds, gardens, and cosmetic landscaping. Large open spaces designed for a variety of activities are also amenities, since they ordinarily do not occur with lower-density housing. Under certain circumstances, shared energy systems can be designed for cluster housing. This can further reduce energy costs beyond the reductions that can be obtained with individual-unit design. Inclusion of all these amenities may not be possible with present economics. Therefore, the appropriate amenities need to be identified for each anticipated group of occupants.

Cluster Concepts and Characteristics

What distinguishes a cluster from other forms of housing? Or more specifically, how is low-rise, medium-density cluster housing different from other forms of housing? The answers to these questions reveal the basic concept behind cluster housing and identify its individual characteristics. A *cluster*, by definition, is a close grouping of elements, or, in this application, a close grouping of housing units. Further, the individual units making up a cluster share common building elements and form common spaces. A cluster describes this sharing in both a physical and social setting. In *Site Planning for Cluster Housing*, by Richard Untermann and Robert Small, cluster housing is described as follows:

> Cluster housing is primarily an urban form that is adaptable to many different community scales. By drawing upon the best of a rich tradition, it has the potential to become the enlightened compromise between conventional suburban and urban housing environments for which so many Americans are searching.[71]

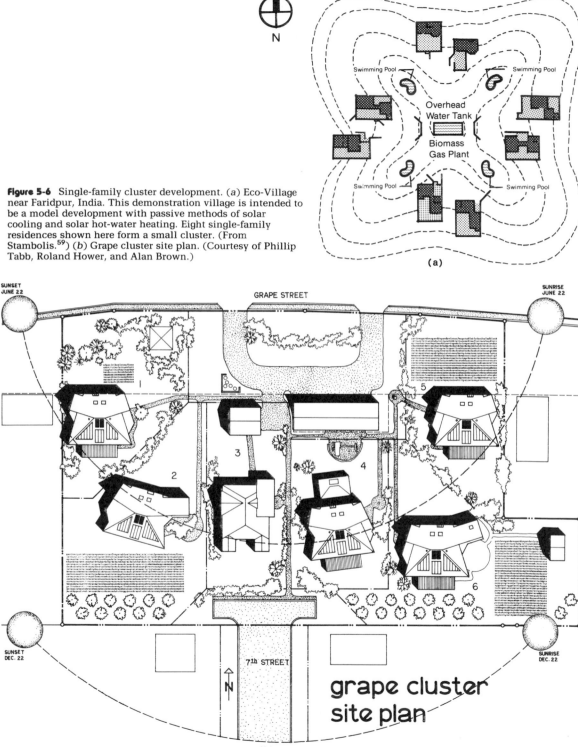

Figure 5-6 Single-family cluster development. (a) Eco-Village near Faridpur, India. This demonstration village is intended to be a model development with passive methods of solar cooling and solar hot-water heating. Eight single-family residences shown here form a small cluster. (From Stambolis.[59]) (b) Grape cluster site plan. (Courtesy of Phillip Tabb, Roland Hower, and Alan Brown.)

Single-family houses can be arranged in a cluster so that there is a shared exterior open space but no shared building elements. High-rise housing has shared building elements—walls, floors, ceilings, etc.—but generally does not form an integral and accessible shared open space. Linear attached housing has attached units that can define an exterior open space. The difference between the open space of the linear form and the cluster form is a function of whether or not the open space is shared, or communal. Of course, with the cluster design, the open space is assumed to be shared.

In physical terms, a cluster is a grouping of attached housing units that simultaneously define a shared exterior open space or series of open spaces. The basic concept of the cluster is expressed by the "doughnut" form, in which all units are linked together and adjoin a center open space. This is a simple, powerful form. Further characteristics of cluster housing include a strong focal point or focal outdoor area, a sense of place with a communal definition of territory, and a sense of entry to the cluster.

Linear housing is generally a simple repetition of housing units. Measures promoting privacy are reinforced—both indoors and outdoors. There are few, if any, communal areas. The linear arrangement, with proper orientation, can be easily adapted to solar heating systems. Active solar energy systems can easily be placed on the roof of each unit, and passive solar energy systems can occupy southern facades. In contrast, the cluster form is more open, creating a center with housing units facing onto the central space. For units with a southern exposure, solar energy systems can easily be integrated. Difficulty arises for those units with poor southern orientation. A major challenge facing energy-responsive cluster housing is the resolution between the single orientation of solar energy systems to the south and multiple orientations of the units to the center. As many as half of the housing units can have poor access to the sun. Figure 5-7a and b illustrates cluster housing versus the linear arrangement of attached housing.

There is a compromise to this apparent conflict of south orientation and central place within the cluster. The solution is an obvious one. The cluster formation can be made with all south-facing units. The open space is formed by slipping or jogging some of the units along the north-south axis, thereby creating a central place. (Refer to Figure 5-5.) With this approach, the orientation requirement of solar energy systems is satisfied in the cluster form. Internal shading may occur. Therefore care should be taken when locating actual passive and active solar energy systems on south facades and roofs. A delicate balance exists between the internal shadows caused by the amount of slipping in plan and the area requirements and positioning of solar energy systems. Under normal circumstances, the balance can be maintained.

Energy-Oriented Cluster Design

Many measures can be taken to make cluster housing more energy-efficient. Energy conservation and solar energy systems applied to individual

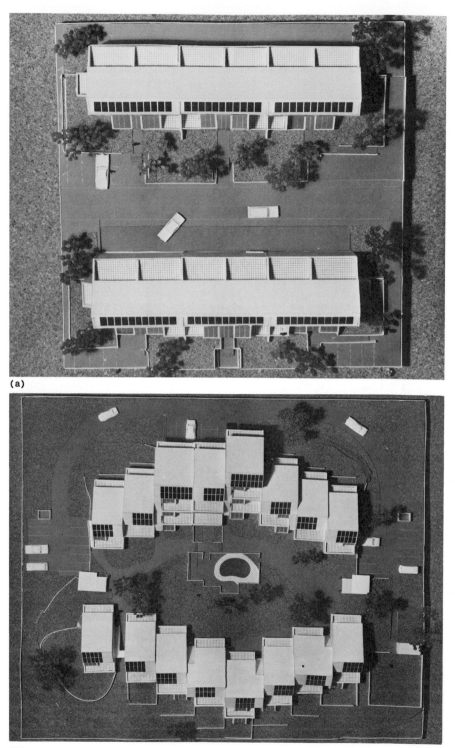

Figure 5-7 (a) Linear housing and (b) cluster housing.

(a)

(b)

Figure 5-8 East-west-oriented cluster, model photograph. In this cluster design, black areas indicate active solar collectors for domestic hot-water heating and gray areas indicate various passive space-heating systems. (Student project, University of Colorado at Boulder, James Elliott.)

units, and shared energy systems applied to multiple units or the entire cluster can greatly reduce consumption. Landscaping can help reduce energy demands while serving needs of privacy and aesthetics. Although the needs of each specific site, building program, and development generate the nature of a cluster project, several generic energy-oriented concepts are worth elucidation. The first concept is the east-west-oriented cluster, the second concept is the north-south-oriented cluster, and the third is the free-oriented cluster, or one with no specific orientation requirements. Descriptions of these approaches follow.

The *east-west-oriented cluster* is the most straightforward concept of the three. As illustrated in Figures 5-8 and 5-9, this concept orients all housing units toward the south. Two crescent-shaped building masses are arranged so that each unit has an integral solar domestic hot-water heating system along with a variety of possibilities for passive space-heating systems—direct-gain, sunspace, or thermal-wall. In the figures, the domestic hot-water collectors are indicated in black tone and the passive collectors are indicated in gray tone. Thermal mass can be located on the floor adjacent to south glazing and in walls along the north-south axis, where it is singly charged. Throughout the day some portion of the central open space is in shadow. This is true for summer as well. Because of the apparent curvature of the cluster in plan, units to the center of each building mass can be taller without causing appreciable solar-access problems. Winter morning and afternoon shadows should, however, be analyzed to determine what the maximum height of the southernmost building can be.

The cluster illustrated in Figures 5-8 through 5-14 is of student design. The two buildings form an east-west-oriented center space. Circulation to each of the housing units and communal areas occurs through a fairly

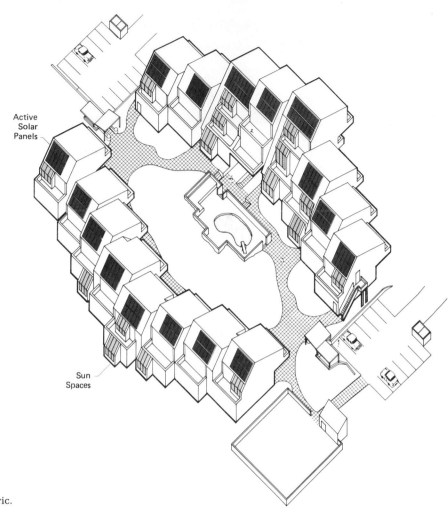

Active
Solar
Panels

Sun
Spaces

Figure 5-9 Cluster isometric.

large hard-surfaced area taking up nearly half of the area of the center space. The adults' and children's swimming pools are surrounded by sunbathing areas and a natural green space with trees for summer shading. Both the green space and the swimming areas are raised approximately 6 feet above the rest of the center space in order to ensure greater solar access. The entries of the center space are delineated by constriction, steps, and additional green space and landscaping. At the southeast corner of the cluster is a community garden for all of the residents, and at the southwest corner is a laundry and a day-care facility with an outdoor children's play area. Parking occurs in three lots located to the east, west, and north of the cluster. The farthest distance from a center unit to parking is 100 feet in this scheme.

In the design for the east-west-oriented cluster, the tallest part of a building is four stories, with housing units stacked on top of one another. To create variety, ground-level units have changes in floor level, and

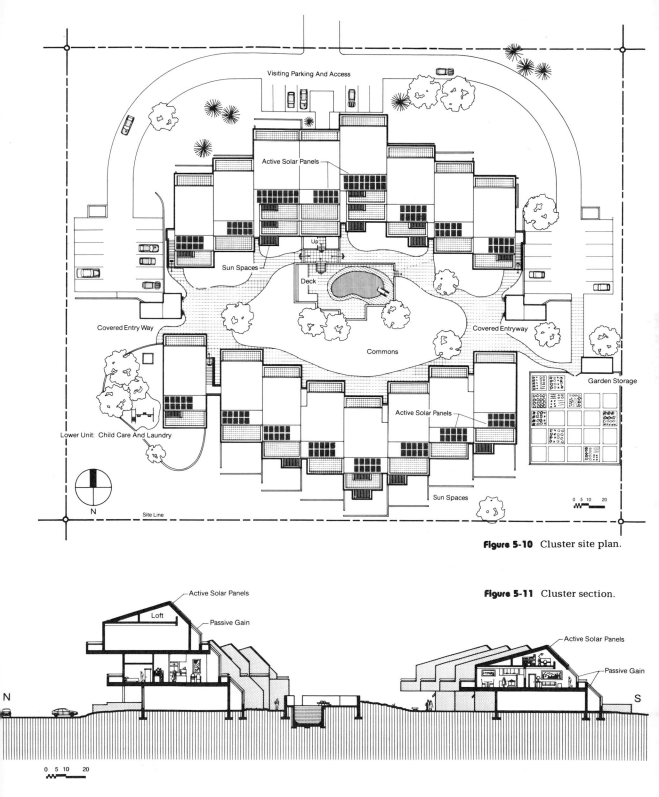

Visiting Parking And Access

Active Solar Panels

Sun Spaces

Up

Deck

Covered Entry Way

Covered Entryway

Commons

Garden Storage

Lower Unit: Child Care And Laundry

Active Solar Panels

Sun Spaces

N

Site Line

0 5 10 20

Figure 5-10 Cluster site plan.

Figure 5-11 Cluster section.

Active Solar Panels

Loft

Passive Gain

Active Solar Panels

Passive Gain

N

S

0 5 10 20

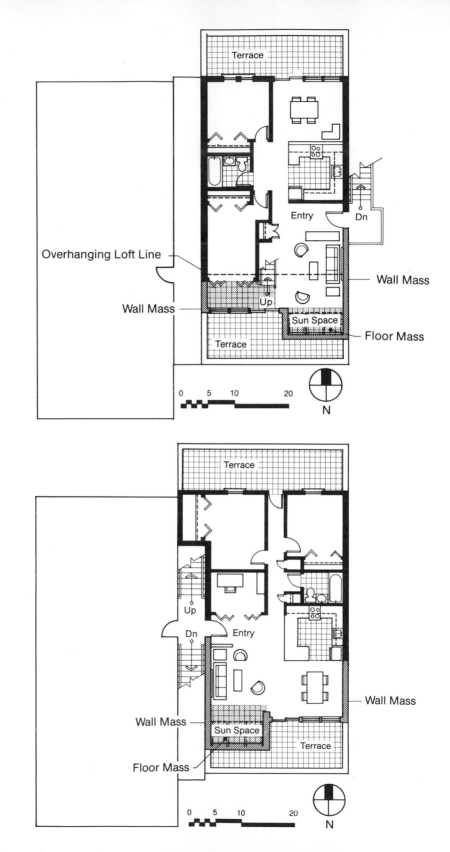

Figure 5-12 South-unit floor plan.

Terrace

Entry

Dn

Overhanging Loft Line

Wall Mass

Wall Mass

Up

Sun Space

Floor Mass

Terrace

0   5   10   20

N

Figure 5-13 North-unit floor plan.

Terrace

Up

Dn   Entry

Wall Mass

Wall Mass

Sun Space

Floor Mass

Terrace

0   5   10   20

N

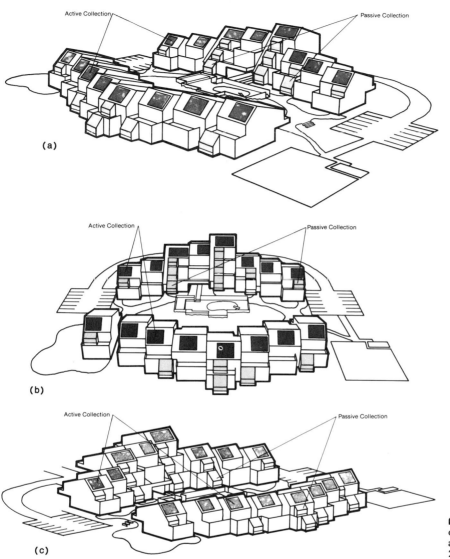

(a)

(b)

(c)

Figure 5-14 Solar-access computer analysis for (a) 9:30 a.m., (b) 12 noon, and (c) 2:30 p.m.

upper units have varying ceiling heights. Density tends to increase toward the center of the buildings. Located on the southwest corner of the northern building is an indoor community space. Many activities can be programmed for this area—day care, indoor recreation, etc. The center of the cluster has a platform and swimming pool. Directly to the south of the pool is a covered multipurpose outdoor space that can serve for many activities. Private outdoor spaces are generally located to the perimeter of the cluster.

The *north-south-oriented cluster* is similar to the east-west-oriented cluster, but it has been rotated 90°. The long axis of the buildings and the central space runs along the north-south axis. With this orientation, several changes occur. The most dramatic change is the north zoning of housing units. Only the two southernmost units receive direct southern sunlight. Therefore, more extensive methods of roof, east-wall, and west-wall solar collection are necessary. Clerestories, roof monitors, and rooftop active solar collection are effective with this cluster design. (Refer to Appendix A for sawtooth collector spacing.) Thermal mass can be located along the north walls of each unit and charged from above. Larger east and west glazings can be applied but should be accompanied by movable insulation and proper window management. Jogging of housing units can open up southern exposures; however, they tend to be shaded half of the day. Shading on the center space is quite different as well. In early morning and late afternoon, shading is extensive. However, during midday hours, there is little to no shading of the center space.

Figures 5-15 and 5-16 illustrate a north-south-oriented cluster. Rather than having varied building heights, that is, one-, two-, three-, and four-story unit masses, the north-south-oriented cluster should have a more constant roof elevation, with rooftop methods of solar collection. Avoid the stacking of units, where roof access is difficult for all housing units, unless some south access is possible. The town house is the most effective housing form in this situation. If solar collection is reduced with this concept, perhaps greater energy-conservation measures should be employed. The swimming pool, garden, and other sunny areas are located to the south of the open center space. A shaded or seasonally covered outdoor space is appropriate to the north of the center space. Parking occurs in small lots on the east and west.

The *free-oriented cluster* can be quite different from the two cluster concepts just discussed. "Free-oriented" means free of orientation constraints other than those that would occur naturally in any project due to the site shape or slope. The energy design approach is twofold: (1) The buildings can be superinsulated, thus reducing the energy demand and the subsequent area requirement of a solar heating system. This helps free up building placement on a site. (2) The solar energy system is detached and possibly remote. In this case the solar energy system should probably be an active air or water space-heating system. Solar thermal storage can be either in one large remote storage tank or in separate storage tanks located within each housing unit. This particular solar energy system concept will be discussed in more detail in the next section of this chapter.

Figures 5-17 and 5-18 illustrate the design of a free-oriented cluster. With the solar energy system separate, the buildings are able to be located to best serve the needs of the cluster. Note that the housing units are in a radial pattern with equal focus to the center space. The solar collector array and storage are remote and located near the "back door" of the cluster. The swimming pool, day-care facility, and other shared areas are located closer to the "front door" of the cluster. Direct solar gain, mainly

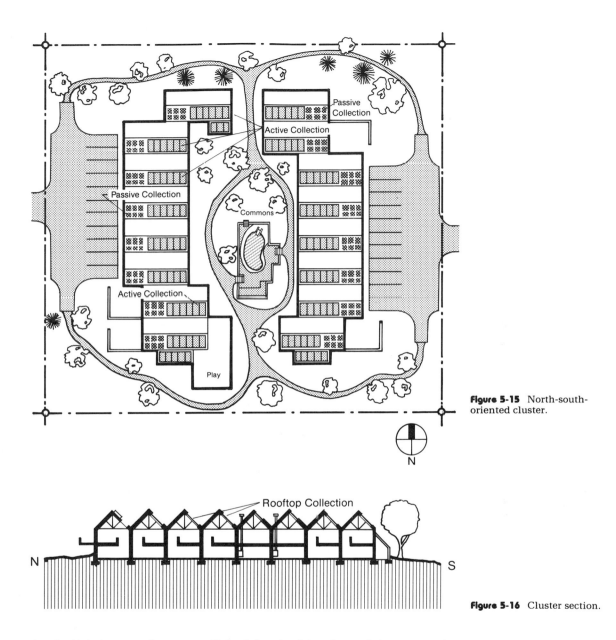

Figure 5-15 North-south-oriented cluster.

Figure 5-16 Cluster section.

for daylighting, can be accomplished, but in this scheme it is not a major design determinant. With this cluster design approach, a tighter, or more densely packed, cluster can be achieved if it is desired.

Another approach that lends itself to the free-orientation concept is earth-sheltered clusters. With these, heavy insulation along with the earth sheltering can dramatically reduce conduction, convection, and infiltration losses and gains. A more constant comfortable temperature range can be maintained within the housing unit year-round. In certain instances, this approach can be conceived as a shared energy-conserva-

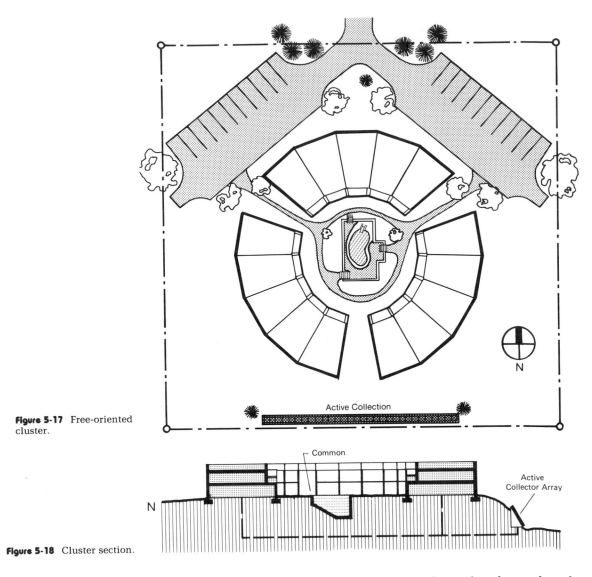

Figure 5-17 Free-oriented cluster.

Active Collection

Common

Active Collector Array

N

Figure 5-18 Cluster section.

tion approach. With the close grouping of housing units, the earthwork is integrated in a unified manner. The cluster comes off as a whole rather than as competing individual elements.

SHARED ENERGY SYSTEMS

One of the greatest challenges for the utilization of solar energy is in shared energy systems. Applied to single buildings, solar energy systems have been proved over and over again. However, few systems have been applied to multiple buildings; that is, few single solar energy systems have been designed to provide space heating and domestic hot-water heating for multiple housing units and numbers of buildings. Many solar

TABLE 5-1
Comparison of Solar Energy Systems

System	Economics	Social Aspects
Direct-gain	Cannot be shared	No benefit
Thermal-wall	Some savings possible	No benefit Potential equity problem
Sunspace	Some savings possible	Shared indoor activity Privacy problem
Superinsulation	Great savings Relatively short payback period	Increase privacy Equity potential
Active air	Savings possible Air-distribution problems descriptive	No benefit Potential equity problem
Active water	Savings probable	No benefit Potential equity problem
Domestic hot-water	Savings probable	No benefit Potential equity problem

energy systems can technically provide energy to a cluster development. The size and number of independent solar energy systems are dependent upon several factors, including cost, location of the systems, distribution, metering, and aesthetics. The principle of the economy of scale seems to apply to energy systems as well. The somewhat contradictory notion of centralizing decentralized renewable energy systems may prove appropriate to the scale of cluster design. The sharing of solar energy systems seems to fall into two areas: (1) *small-scale sharing*—between two or three units—and (2) *entire-cluster sharing*—among all the units and buildings.

Small-Scale Sharing

Small-scale sharing of energy systems is generally limited to several housing units. At this smaller scale, the sharing of some energy systems is quite possible. Three considerations are important: the technical feasibility of the energy system, the economics involved, and the social ramifications of the use of shared energy systems. Each of these considerations needs to be addressed in order to create a successful solution. If, for example, there are ambiguities associated with the metering of conventional fuels that may be tied into the function of the shared energy system, problems can arise. Therefore, clear methods directing an equitable use of the shared systems and conventional fuels must be provided.

Taking each of the generic solar technologies into consideration, the relative merits of shared energy systems can be analyzed. The solar technologies are passive space-heating systems (direct-gain distributed-mass, thermal-wall, sunspace, and superinsulation), active space-heating systems (water and air), and active hot-water heating systems. Table 5-1 compares these solar energy systems according to probable economic benefits and social amenities and potential difficulties. Both the eco-

Figure 5-19 Student dormitories at Colorado School of Mines, Golden, Colorado. John Anderson Associates Architects.

nomic and social aspects are generalized and depend upon specific sites, users, and user interactions. As can be seen in the table, some of the solar energy systems lend themselves to sharing while others do not.

It is quite evident, through examining the information in the table, that the direct-gain and thermal-wall passive systems are not conducive to sharing. They are dependent upon and restricted to specific thermal zones. Each needs a fairly direct relationship to individual unit facades. Because of possible thermal distribution problems and interface conflicts with conventional heating systems, the active air system may also not be feasible. Duct sizes could be very large. The remaining four solar energy systems—the sunspace, superinsulation, active water, and domestic hot-water heating systems—offer possibilities for shared or consolidated energy systems. The economic and energy-conservation benefits to sharing are simple: less repetition of systems and components, less labor and materials, more efficiency, and less overall energy loss.

An active solar heating system can be designed to provide both shared space heat and domestic hot water. The collector array can be made of many individual solar collectors that are tied together; one well-insulated thermal storage tank can be used for space heating, and one can be used for domestic hot-water preheating. The domestic hot-water heating system would be similar to the space-heating one, yet somewhat smaller due to the smaller load. At the front of the distribution systems there would be individual meters and pumps to deliver the solar-heated hot water to baseboard radiators and service hot-water outlets located throughout each unit. With this system design, the collector array can be located anywhere on or near the building. Any number of arrays can be integrated into the building form as well. Collectors are not directly related to housing units. Refer to Figure 5-19.

A passive solar heating system can be designed to provide shared space heat. It can be an isolated direct-gain type of passive system. In this type of system, solar heat is collected in the sunspace or thermal wall and then

TABLE 5-2
Comparison of Renewable Energy Systems

System	Economics	Potential Problems
Active solar	Savings in collection and storage cycles	Glare Aesthetics Little peak load capacity Area-intensive
Photovoltaic	High initial cost Energy-intensive processes	Glare Aesthetics Dependent upon direct sunlight access Area-intensive
Wind machines	Often productive for peak energy demand	Aesthetics Storage
Geothermal	Often productive 24 hours a day	Limited access
Hydroelectric	When operating, can be productive 24 hours a day	Limited access Costly for this scale
Passive solar	Added functional area Center could be an atrium	Costly Inefficient Area-intensive Distribution problems

distributed directly to various housing units or delivered to one or more storage beds, which in turn deliver heat to various housing units. The entry, common greenhouse, laundry, and party room can all be located in the sunspace. This space can also be used as an area for a variety of communal activities, especially gardening, or as an enclosed semipublic space that acts as a transition between the outdoor public spaces of the cluster center and the private housing units.

Whole-Cluster Shared Systems

Whole-cluster shared energy systems are not limited to just a few housing units; they can encompass the entire cluster. One or more solar energy systems can provide either space heating and/or domestic hot-water heating for each housing unit within the cluster. While many smaller shared systems could in fact provide solar energy to the whole cluster, this section will discuss using a variety of single systems for the entire cluster. In other words, the solar energy systems considered here will be centrally oriented systems of renewable energy sources. The cost savings occur primarily in the collection and storage cycles. The distribution of solar energy to the individual units is accomplished through conventional means and, therefore, has little, if any, economic benefit.

The types of renewable energy systems capable of providing energy to the entire cluster are limited to a few, primarily active. They include an active liquid solar heating system, a photovoltaic electricity-generating system, and wind electricity-generating machines. Other types of renewable energy systems may be appropriate if the energy sources are available, such as geothermal energy systems or hydroelectric systems. Table 5-2 compares these systems according to potential economic benefit and

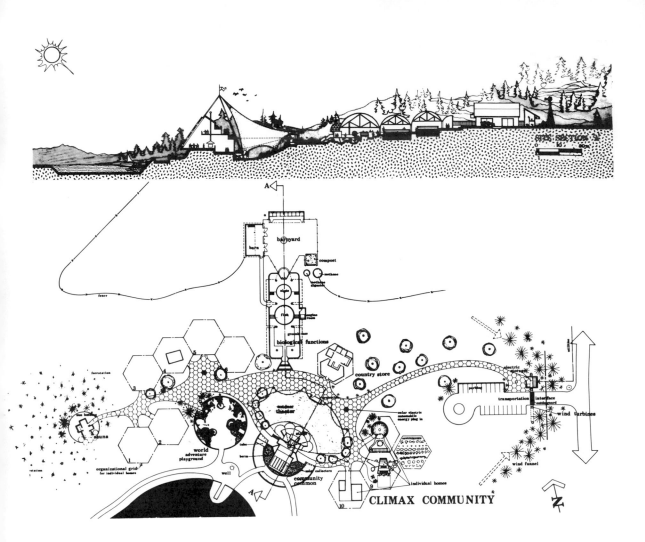

FIGURE 5-20a Community site plan. The following outline identifies the major design features of this modest solution.

1. Site analysis, phased design, and program combine to form the larger environment.
   a. Minimal auto impact.
   b. Electric carts and bike paths throughout site.
   c. Rainwater holding pond connects to lake and acts as community focus.
   d. Community located for optimum solar gain.
   e. Wind funneled to turbines at site entry.
   f. Density appropriate to semirural location.
2. The community is composed of sociological, biological, and individual functions.
   a. Community common and theater located at center of development.
   b. World adventure playground and child-development center adjacent to community common.
   c. Community sauna in forestation area.
   d. Holding pond irrigates growing fields, which support animals pastured downwind from community.
   e. Biological functions described in diagram produce food for residents, with surplus marketed in the country store.
   f. Hexagonal organizational grid suggests the private domain for ten families.
3. Life-support units within the community satisfy individual needs for privacy, energy, and food.
   a. Within the hexagonal grid is an area to support individual housing, energy collection, growing, and recreation.
   b. Each family designs its own environment.
   c. Adjacent hexagons offer different levels of sharing.
   d. Units are designed to optimize energy collection and conservation.
   e. Individual greenhouse integral with kitchen functions.

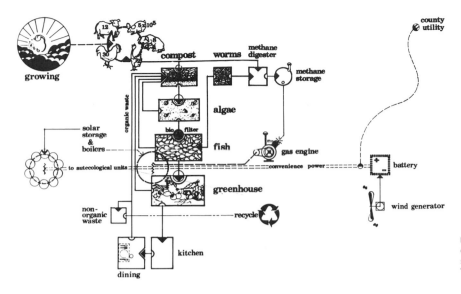

Figure 5-20b Synecological diagram of shared energy systems. (Courtesy of Phillip Tabb Architects.)

potential problems associated with their use at this level. A passive space-heating system is also included.

The choice of the best energy system is not obvious. Not every development site is going to be adjacent to a river or stream where hydroturbines may be placed. Nor is a development commonly going to be located near geothermal reservoirs, which most often are found along areas of volcanic activity. Many areas of the United States report extremely high winds. Often these winds are not sustained and therefore cannot economically support wind turbines. A large passive atrium may seem to be very attractive, but it is costly and area-intensive. Photovoltaic arrays are very expensive and haven't really been proved on the commercial market. The local conditions surrounding a particular cluster development are extremely important in determining the most appropriate approach to a shared energy system. Should a site have access to sustaining amounts of any of these renewable energy sources, a shared energy system could be an exciting addition to a development project.

The energy systems planned for the example in Figure 5-20a and b represent a variety of alternative energy approaches to a small cluster. The design is for a small self-sufficient community of ten families—a population of approximately forty to fifty people. Although the design is not economically realistic, it does demonstrate many of the types of shared energy systems. Because of the level of sharing and the extent to which community facilities have been planned, the development can be planned for a much larger number of families (this would be economically beneficial). Ten hexagonal single-family plots are organized around a community center. The designs for each of the homes are individualized. Each single-family residence is designed with its own solar energy system for space heating and domestic hot-water heating. There is a community center with shared kitchen, dining, day-care, and meeting areas and an outdoor theater. In addition to solar heating, the community systems include methane gas production, wind-machine electricity generation, electric cars, and a community greenhouse for food production and fish farming. There are also growing fields, shared gardens, farm animals, and grazing fields. A meditation space is provided in a secluded place.

The shared energy systems that seem to have the greatest promise for space heating and the broadest application for more common whole-cluster development at the present are the active solar space-heating and hot-water heating systems. Although the energy from these systems is lower-grade in nature (as opposed to electricity or steam produced by some of the other renewable systems), they are generally available and have been proved to work. The solar collectors required for the system can either be located on the roofs of the buildings or be isolated on the site. Thermal storage can be accomplished in one or more insulated tanks. In some instances, annual storage, where water is charged throughout the year, can be incorporated with this system. Refer to Figure 5-19.

The rationale behind the use of annual storage is simple. Once a solar system is in place, it has the potential to collect solar energy year-round. Annual storage is potentially more cost-effective than storage on a seasonal basis. Thermal storage for winter use could begin in summer. By incorporating large reservoirs of water, generally at ground temperature, a solar thermal system can charge the water or working fluid and slowly raise its temperature. In fall, winter, and spring this water can be coupled to a variety of conventional heating and hot-water systems and possibly boosted in temperature. Instead of using the initial ground-water temperatures of around 45 to 55°F, a system such as this could provide initial temperatures of 80 to 95°F. Disadvantages of this approach include a high initial cost for the relatively large storage volume and possible placement problems on a given site. The annual storage system can also be seen in Figure 5-20.

The use of solar ponds is increasing. It is limited, of course, by specific site and area constraints. There are several advantages to the solar pond in the event the site, area, and cost constraints are reduced. Normally solar collection is area-intensive; a solar pond can be made very large with negligible amounts of construction materials. The solar pond has its own "built-in" storage system. A solar pond, according to Dr. Harry Tabor of the Scientific Research Foundation in Jerusalem, is normally 2 meters (approximately 7 feet) deep. Deeper ponds can be used for seasonal storage. A limitation to the use of solar ponds is the problem of finding feasible locations. Since solar ponds are horizontal, they should be located near the equator—usually no farther away than latitudes of 40° north or south. The practical utilization of solar ponds is most probably going to be in warmer, third-world countries.

CLONING CLUSTERS

The concept of cloning often has negative connotations. Somehow the repetition or multiplication of elements is seen to take away from identity or individuality. This is not necessarily true, especially at the planning level, where the concept of cloning can be useful. Many of the cluster designs illustrated in this chapter have been planned with similar populations of approximately 75 inhabitants or 26 housing units in mind. If a development project is programmed for 78 units or as many as 234 units, the planning problem is more complex. By cloning the first cluster of 26 units, the problem can be simplified to the planning of three clusters for the 78 units and nine clusters for the 234 units. At this level creative planning takes into consideration the relationships among clusters and the even larger scale of sharing and amenities that is possible. The next two sections will focus on two levels of cloning: the smaller and larger arrangements of clusters.

Small Arrangement of Clusters

Several physical planning determinants are important in the siting of solar clusters. The site-planning determinants include planning for solar access, planning for the automobile, designing the open spaces, and providing the appropriate amenities. Clusters can be planned as isolated, independent subneighborhoods, or they can be planned for greater levels of interaction and dependency. Many of the cluster needs that were discussed earlier in this chapter hold true for the cloning of clusters. Economy, privacy, and convenience should especially be considered in relation to the site-planning determinants previously mentioned. The cluster design for Marin Solar Village in northern California clearly illustrates these planning determinants.

The Marin Solar Village is a planned self-sufficient community near San Francisco. The project was designed and planned by Van der Ryn, Calthorpe & Partners. This new town is planned for nearly 2000 dwellings, most of which are in the form of attached, east-west-oriented clusters. Typically a cluster is made of 32 dwellings with semienclosed central spaces for shared activities. Slight variations occur from one cluster to another. For example, periodically clusters have additional pedestrian access to the center space through the north or south buildings. Often several clusters will form an even larger community space. Refer to Figure 5-21, which illustrates a typical grouping of solar clusters for this project. Parking for both visitors and residents of the clusters is located no more than 400 feet from a housing unit. This is fairly good for multifamily housing, but 100 feet would be more convenient. Dimensionally, a cluster is in the range of 200 feet in the east-west direction by 150 feet in the north-south direction.

The solar energy systems for the clusters are simply direct-gain windows, clerestories, and sunspaces arranged on the south facade of each

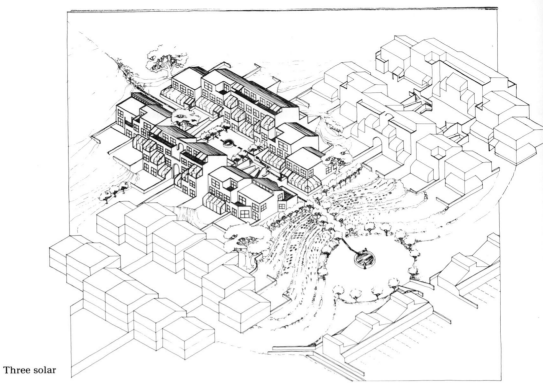

Figure 5-21 Three solar clusters.

building for space heating. A few of the units have active solar collection for domestic hot-water heating. Each housing unit has either a sunspace or ample vertical south glazing. Units located above ground level have southern decks or balconies, while ground-level units have private gardens. Where two or more clusters join, either private gardens adjoin or the clusters form a public open space. These open spaces are typically used as parking and recreation areas or park land. A larger-scale entry could be suggested. Cloning in this example has been accomplished in a sensitive way. The spaces between clusters can facilitate privacy, and they can easily break the monotony of simple repetition.

Larger Cloning Arrangements

For cloning larger numbers of clusters, two basic patterns exist. The first pattern is the rather static or consistent repetition of the cluster module with its needs, forms, energy systems, and internal center space. While the cluster remains constant, it is assumed that individual styling will occur at the unit level. The module functions to a certain degree as an independent organism; each cluster is relatively complete. In the second pattern there is also repetition, but community and support facilities are individually designed. The greater the number of clusters, the larger is the economic and social base for cooperation and use. When both come

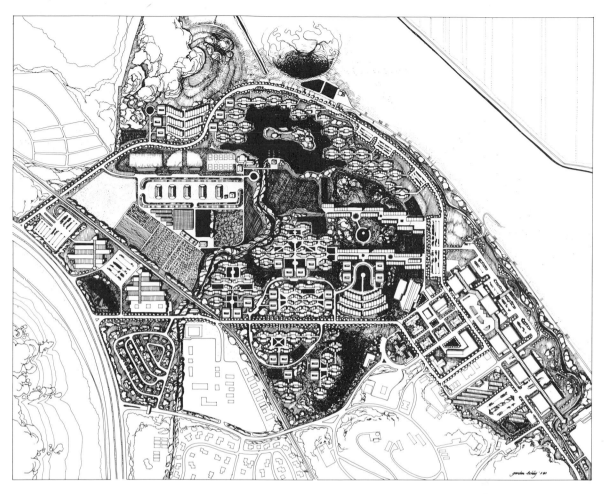

Figure 5-22 Marin Solar
Village. (Courtesy of Van der
Ryn, Calthorpe & Partners.)

together, a neighborhood or even a community can be formed, depending
upon the scale of the development. The broader implication of large numbers of clusters suggests the need for care at the planning level in order
to balance the many potentials and problems associated with any large-
scale development. Figure 5-22 illustrates a pattern of cloned clusters in
Marin Solar Village.

There are many advantages possible at the larger development level.
The first advantage is the convenience related to greater social contact
and the variety of interactions possible. Activities for all age groups can
take place within walking distance. Aesthetics is another advantage that
can come from sensitive design and planning. Automobiles can be relatively remote from communal activities, and landscaping can enhance
the everyday experience of both central and transitional spaces. The network of pedestrian and bicycle paths can provide pleasant circulation,
discovery, and recreation. Safety is another spin-off of the isolation of the

automobiles. The last advantage, and perhaps the most interesting in relation to the subject of this book, is the use of renewable energy systems. It is possible to integrate a variety of renewable energy systems into the larger cluster developments. These systems not only can provide needed energy but can serve as a symbol demonstrating a more natural and self-sufficient way of life.

SUMMARY

In meeting the new housing needs of the near future, multifamily housing will certainly have its place. The challenge to make this higher-density form of housing more humane in the light of growing energy and construction costs is the key to its success. The low-rise, medium-density cluster is an attractive answer to this challenge. The energy and cost savings associated with cluster development are quite evident. And perhaps the greatest benefits of cluster housing will be the social or community ones—notwithstanding the stresses on privacy—the ones that foster group sharing and growth.

NEIGHBORHOOD 6
PLANNING

*If we assume that urban dwellers could, under
proper conditions, walk for up to thirty minutes
in order to go to their jobs, their theatres, their
markets, etc., then this distance of a mile and a
half appears to be the maximum distance
which can be allowed for a sector under the full
control of man.* CONSTANTINOS A. DOXIADIS[22]

The *neighborhood* is the geographic area within which residents conveniently share the common services, support systems, and facilities needed in the vicinity of their dwellings. The neighborhood is the scale of measurement for linking comprehensive community plans with people and the areas they inhabit. Composed of subdivisions, planned unit developments, single-lot developments, and many community facilities, the neighborhood is useful in guiding new suburban residential land use along with the redevelopment of central city areas. In the creation of neighborhoods, planning-level decisions are made, many of which have an effect upon the potential source and conservation of renewable energy, especially solar energy.

This chapter is the third in the sequence of discussion of residential planning scales. The scales of shelter and cluster design primarily deal with housing forms; the neighborhood is one context within which these forms exist. Both solar heating and transportation are addressed at this scale. The age of abundant automobiles has made it possible for most adults to travel fairly long distances in order to satisfy a myriad of needs. Work, grocery shopping, specialty shopping, recreation, and many other daily needs and services have been made accessible because of the automobile and the availability of inexpensive fossil fuels. In many communities the neighborhood is the area of city life that has suffered the most. It no longer performs the intrinsic function of satisfying the basic needs of its residents. At best, it is often perceived as a picturesque reminder of the past.

The concept of neighborhood planning is growing stronger in some communities as the potential for energy savings is seen along with other economic and environmental advantages. A strong neighborhood functions like a local community, and it can be planned to be relatively self-sufficient. The methods of energy conservation at this scale are more likely to be found with planning forms rather than architectural forms. One of the purposes of this chapter is to examine energy-responsive planning techniques applicable to the *subneighborhood*—city blocks, planned unit developments, and mixed-use developments, all of which occur within the neighborhood—as they are the legal and physical vehicles for actually creating neighborhoods and small communities. Another purpose of this chapter is to explore both new and old *neighborhood planning concepts*—their structure and form in creating a unified context for energy conservation. This chapter presents the subneighborhood and the neighborhood as links between the particular concerns associated with multiple buildings and the comprehensive needs of the whole.

SUBNEIGHBORHOODS

Although the neighborhood concept is valid as an overall planning tool, perhaps more important is the creation of subneighborhoods and appropriately designed planned unit developments and city blocks. The planned unit development is becoming a popular vehicle for actualizing many innovative planning concepts. Large subdivisions like Levittown and large new communities like Brazilia conceived two decades ago are no longer accepted planning schemes. They have proved to be too large and inflexible, and they do not allow for enough organic growth. The planned unit development is smaller and more contained, and it is organized around a process with local governmental discretion. The ability to introduce mixed uses within the planned unit development has also made it more popular. A planned unit development can comprise a city block or several blocks, depending upon size, density, and existing composition of buildings. Energy planning applied to the planned unit development, mixed-use development, and city block can make the subneighborhood a much more important planning scale.

Figure 6-1 is an illustration of a grouping of solar-oriented single-family houses designed for U.S. Homes by architect Dennis Holloway. The informal siting of the houses, with varied orientation, adds to the neighborhood setting. The design is composed of houses derived from the traditional "salt box," which has large window areas to the south for passive collection and a large roof area to the north for protection from cold weather. This two-story design also features a Trombe wall on the lower level. The project is part of the National Association of Home Builders (NAHB) and Solar Energy Research Institute (SERI) demonstration program.

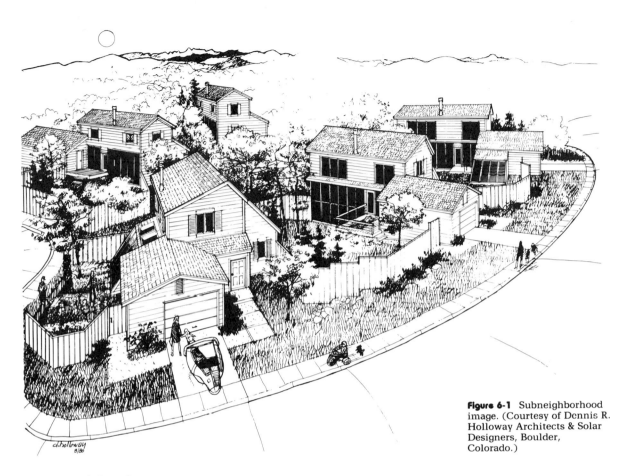

Figure 6-1 Subneighborhood image. (Courtesy of Dennis R. Holloway Architects & Solar Designers, Boulder, Colorado.)

Planned Unit Development

A *planned unit development* (PUD) is a land-development concept with allowable land-use mixes, including housing at varying densities and commercial, open-space, and natural features. Several PUDs can be grouped to form a neighborhood. The PUD discards land-use districting for a more open process allowing the development of various types and sizes of parcels ranging from small clusters to entire new communities. A transfer occurs here from the detailed concerns for individual and family needs for shelter to concern for the needs of the neighborhood and community. Energy-conservation techniques and solar energy systems applied to this scale of development offer exciting new possibilities for both saving energy and evolving community form.

A PUD can be conceived, built, and occupied within a couple of years. Although as few as 5 or 6 housing units can comprise a planned unit development, the accepted number is closer to 50 or more units. With higher-density projects the figure can even be higher, 200 to 300 units. At the larger number of units, more amenities and community services can be built into a scheme because of the improved economics. Certainly planning for solar access is important at this scale of development. Plan-

Figure 6-2 Photograph of Wildwood solar development. (Courtesy of Milburn-Sparn Energy Architects, Inc., Boulder, Colorado.)

ning for the diversified energy requirements of various mixed uses is also important. In an article entitled "Beyond Solar: Design for Sustainable Communities," Peter Calthorpe and Susan Benson put into perspective energy savings associated with solar technologies and energy savings associated at the planning level.

> Quite simply, adding passive solar systems to American homes does save energy inside but does not take into consideration the accompanying land use, infrastructure costs, and transportation demands. In truth, the form and density of housing, the land use patterns, and the resulting transportation systems have a much greater potential for energy savings than any solar applications.[11]

Two overall energy planning objectives are appropriate for PUDs; they concern *building energy consumption* and *transportation energy consumption.* The first objective is to protect or allow for climate-responsive architectural design for individual buildings within the PUD. Energy savings can occur with reduced building energy demands for heating or cooling, depending upon the specific climate. The second objective is to reduce automobile-intensive design. Energy savings can occur with reduced resource consumption, especially gasoline and oil consumption. This includes designing alternative means of movement—pedestrian walks, bicycle paths, and public transportation—mixed-use development, and more efficient systems of roads, streets, and utilities.

To have climate-responsive building design within the PUD, there must be a matching of program or economic objectives with the specific site constraints, such as varying topography, nonsouth orientations, and obstructions to solar access caused by adjacent buildings or objects. This can often be a difficult problem. Traditionally the economic parameters have dominated energy-oriented ones, and as a result, PUDs have been realized with little consideration for energy or at least solar energy. At

Figure 6-3 Typical house.

best, energy has been considered in a piecemeal way. For new developments both the economic parameters and the limitations of solar energy will need careful examination. Both short-term objectives and long-term goals may figure into this relationship. The PUD shown in Figure 6-2 is a blend between these two planning determinants. Although high solar fractions are not anticipated, this is a good attempt at a solar-tempered housing complex.

The figure illustrates a solar-oriented PUD in Boulder, Colorado. The project was designed by Milburn-Sparn Energy Architects. Planned along an east-west-oriented stream, thirty-three single-family detached and multifamily housing forms create a fairly compact subneighborhood within 6 acres. In fact, many of the homes are sited several feet from one another. The overall density for the development is five units per acre; this includes a fair amount of space allocated to open space along the stream. Houses are either oriented to the stream or sited so as to benefit from a view of the mountains to the southwest and solar access to the south. Lot sizes for the houses vary from about 3500 to 4000 square feet. Each single-family detached house has a finished floor area of 900 square feet with a 300-square-foot unfinished basement. The cost of the housing varies from $75,000 to $110,000 (1983 dollars).

The dominant solar facade is associated with either the front or the side of the typical house. This allows for a greater variety of house orientations. In this example the arrangement of houses resembles a small village. There is a nice spatial feeling to the development even though the houses are quite ordinary. The orientation to the sun appears secondary. The houses are two-story with a basement and usable attic space. Refer to Figure 6-3, which is a photograph of a typical house in the develop-

Figure 6-4 Energy strategies for a city block. Urban Survival Shelter. (Student project, University of Colorado at Boulder, Paul Flehmer, Bruce Flynn, Colin Shimokawa, and Russel Watson.)

ment. Solar access is provided to at least the roof and upper stories of all housing units. This PUD does not have mixed uses but is located near a small commercial shopping center within walking distance.

Block Development

Existing, revitalized, and new city blocks offer opportunities for energy savings. However, the measures in creating these savings are not always easily realized. Most city blocks are composed of many privately owned parcels, and therefore, it may be more difficult to initiate uniform change. In a new block, houses or clusters may be constructed over a period of 5 years. Home owners most likely will be concerned with their own individual property. They may not be able or willing to put effort and money into neighborhood or block improvements. This is not to say that block improvements should be discouraged. They do occur; however, in the past they have been accomplished by forward-thinking individuals, families, and groups in unique situations. Key to the success of block improvement is cooperation.

Areas in which residents can become more self-sufficient at the city-block level include energy, food, recreation, and resource recycling. Although much can be accomplished with each of these levels, the degree of self-sufficiency is dependent upon the amount of resources, time, and money that each resident is willing to contribute to the block or community. At lower densities, city blocks can incorporate common areas for energy collection, food production, and recreational activities, which could reduce automobile travel. The low-density city block typically is bordered by streets on all sides, has an alley, and has housing primarily composed of single-family detached units. Street sides of land parcels tend toward more formality and simplicity, and they form a transition from public to private territory. Housing facades facing the street tend to follow this pattern. Alley sides of land parcels and housing facades and the alley itself tend toward more informality, and consequently, energy production, gardens, waste recycling, and shared space are most often found there. This is also true to a certain extent for parcels that adjoin open spaces, such as the single-family houses in the subneighborhood example shown in Figures 6-2 and 6-3.

Figure 6-4 illustrates a design for an entire urban city block. The design, entitled Urban Survival Shelter, provides for housing (300 units), professional office space, commercial and retail space, parking, energy collection, and food production. Parking is located below grade; the professional offices are located to the south in the three-story structure; housing is to the northeast and northwest; commercial and retail spaces are located on the ground level; food production occurs in greenhouses located on the roof of the residential section; and energy collection occurs on the roof, where six line-focusing collectors, or stationary reflector tracking absorbers (SRTA), are mounted. Designed by a group of graduate students at the University of Colorado at Boulder, the project was a winner in the National Student Competition in Energy-Conscious Design in the spring of 1976.

NEW NEIGHBORHOODS

In the United States approximately two-thirds of all energy consumption occurs in the building sector (residential and commercial buildings) and in the transportation sector. This is a considerable portion. Improved neighborhood planning could reduce energy needs for both buildings and transportation. Much can be done at the individual building level, while many new concepts can be applied at the neighborhood level. Solar access can be ensured; shared energy systems can be employed; and neighborhoods can be planned with diversity of densities, land uses, and transportation systems. Additionally, neighborhood gardens can augment the larger agribusinesses or supermarkets with locally grown fruits, vegetables, and flowers. The potential energy savings at the neighborhood level are virtually untapped.

The neighborhood has undergone a great deal of change. Through the end of the Second World War, neighborhood design was a strong planning method. With the advent of the popular use of the automobile and the postwar baby boom, neighborhood design declined. Larger, more spread-out suburbs became the accepted norm for residential planning. Regional shopping centers drove away neighborhood shops; school busing diminished the importance of the elementary school as a neighborhood element; and streets gave way to creative open spaces. With the awareness of the need for energy-conscious design and planning, the neighborhood once again is being seen as an essential component of residential settlement.

Several factors are important in creating an identifiable, energy-oriented neighborhood. First, a neighborhood is a residential grouping that in some way is set apart from the natural environment or other areas of the built environment. Land uses usually accompanying the residential grouping include commercial spaces, limited small-scale institutional facilities, open space for recreation, and other special-use areas. Usually a neighborhood has distinct boundaries or edges—either natural, human-made, or both. Second, a neighborhood has movement systems organized for both pedestrians and automobiles. The direction of these systems can impact the placement and orientation of lots—important factors in solar energy planning. Third, a neighborhood is determined by the service area of certain community facilities, such as an elementary school and a convenience shopping center. Many new neighborhoods may have community gardens and possibly community energy systems. Fourth, a solar-oriented neighborhood needs solar-access protection with adequate zoning. Shading from buildings as well as trees can cause problems. Fifth, a neighborhood has a utility and energy network. Sixth, a neighborhood has character, and this character lends a sense of place.

These six factors of a neighborhood will be described in relation to the solar energy planning and design concepts previously mentioned in this book. Much has previously been written on energy conservation and solar energy utilization for single buildings. Although these energy technologies are going to change and evolve, the transformations are not likely to present major surprises or design problems that have not already been described. Planning considerations at the neighborhood level will be the major focus of this section of the chapter as there is a shift in emphasis from energy technologies and architecture to neighborhood planning systems, especially *land use, movement, community services, solar-access protection, utilities,* and *character.* Because of the complexity of neighborhood planning, only one design can be considered here. The low- to medium-density neighborhood will be examined because this density can easily be planned for solar energy use.

Land-Use Considerations for Neighborhood Planning

Careful application of energy and land-use concepts for the neighborhood can help make it more self-sufficient. These concepts can suggest the quantity, density, and mix of housing types along with the many possible support facilities and systems. Creating a homogeneous composition of shelter forms interwoven with a diversity of community forms and services is key in developing an energy-efficient neighborhood plan. Comprehensive neighborhood planning may better be accomplished by comparing the activities and functional needs of the neighborhood to the physical characteristics associated with the neighborhood form—the bulk land use, the boundary or edge, and special places. Many land-use patterns coming from this comparison are quite obvious, as can be seen in Figure 6-5.

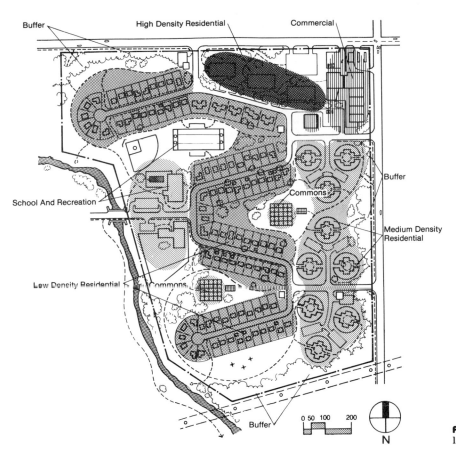

Buffer

High Density Residential

Commercial

Buffer

School And Recreation

Commons

Buffer

Medium Density
Residential

Low Density Residential

Commons

0 50 100   200

N

Buffer

Figure 6-5 Neighborhood
land-use patterns.

The bulk land use, or predominant land-use pattern, of the neighborhood is housing. Single-family detached and attached, multifamily, and cluster housing make up the majority of housing forms at the low to medium density. This does not necessarily exclude the occasional high-rise form. However, the high rise can be the exception. The neighborhood design illustrated in Figure 6-5 is planned for an average density of between 8 to 10 units per acre. This includes approximately 86 single-family detached residences, 12 clustered single-family residences, 15 triplex units, 234 multifamily residences, and 150 high-rise residences for a total of 497 units of housing. This is an approximate population of 1392 at 2.8 persons per family. There are several relationships of the various densities of housing. The different forms can be fully integrated or zoned separately or both. Important influences to the bulk land use are the patterns of streets, paths, and open spaces.

Boundary land-use patterns vary depending upon the activities and land uses adjacent to each neighborhood edge. Typical of neighborhood boundaries are natural features such as streams, forests, hills, or other open spaces, and artificial features such as major streets, large utility lines, legal boundaries, or community facilities. These boundaries can be either positive, neutral, or negative. The type of neighborhood boundary land-use response is, therefore, site-specific or edge-specific. In Figure 6-5 the neighborhood is bounded by major streets to the north and east, by a city utility easement and political boundary to the south, by a stream to the southwest, and by open space with topographical changes to the west. For the purpose of this example, the higher-density housing forms are located near the street boundary edges on the north and east. A landscaped buffer strip actually separates the major street and these housing forms. Single-family plots and single-family clusters are adjacent to the more hospitable open spaces to the south, southwest, and west.

Special land-use patterns are more dispersed throughout the neighborhood. They include open space for community gardens, possible space for on-site energy generation, subneighborhood special-use features such as day-care facilities, and indoor and outdoor recreation areas. Each neighborhood is unique and may develop a different set of special features. In Figure 6-5 the special-use features also include a pedestrian network, which will be explained in more detail in the next section; two neighborhood waste-recycling centers; and a solar-powered neighborhood recreation center designed for indoor recreation, meetings, and a variety of other activities. The incorporation of these kinds of features can add to the character and function of the neighborhood and can add to its energy independence. They also add to the individual unit or property value.

Movement-Systems Considerations for Neighborhood Planning

The average American travels approximately 13,000 miles a year by automobile. The energy required to travel this distance is greater than that required to heat or cool the average house for a year. As a result, movement systems are terribly important in reducing energy consumption and in establishing the overall order of a neighborhood. They can impact neighborhood function and character. This is especially true if streets and roads are an integral part of the neighborhood fabric. Street patterns (not actual traffic) have two major effects on land use. The first effect is the influence on residential lot configuration, including both size and shape. The second effect is on lot orientation. Lot size, shape, and orientation are somewhat limited if solar access is required. The street pattern in Figure 6-6 is a zigzag pattern along the north-south axis with streets within the neighborhood oriented 15° off the east-west axis. Lots are perpendicular to the streets. This pattern allows for variety of housing-unit orientations within the limitations of solar energy collection and breaks up the linearity of the grid.

The street pattern within the neighborhood is served by four major entries: one to the north, two to the east, and one to the west over the stream. These entries are located a sufficient distance from the intersection to eliminate traffic congestion. These through streets form a pattern of various residential densities and open spaces. Two of the streets are terminated by culs-de-sac with clustered housing. Between the units of

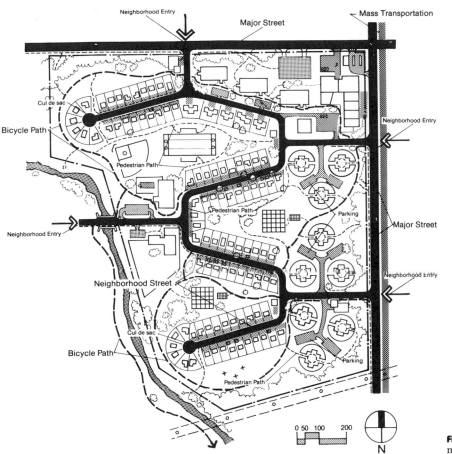

Cul de sac

Bicycle Path

Pedestrian Path

Neighborhood Entry

Pedestrian Path

Parking

Neighborhood Entry

Major Street

Neighborhood Street

Neighborhood Entry

Cul de sac

Bicycle Path

Pedestrian Path

Parking

0 50 100   200

N

Figure 6-6 Neighborhood movement patterns.

detached single-family housing, either pedestrian open space, as shown in the example, or alleys can occur. The form of the streets, alleys, and culs-de-sac need not conform rigidly to the cardinal grid; curved forms are acceptable as long as they do not create difficult lot shapes and orientations. Parking areas can be located along streets, or if off-street parking is desired, they can be located in lots or garages. At an average of 1.5 parking places per unit of housing, there is need for nearly 745 parking places. This is substantial and represents a fairly large land use. The area needed to accommodate these 745 parking places is nearly 7 acres.

Subdivision regulations and PUD requirements, particularly where single-lot developments are concerned, require that lot lines parallel street rights-of-way. Depending upon the size of the lot and the subsequent buildable area of the lot, this requirement can affect the long axis of a house and the house's capacity for economically integrated solar collection devices. This situation is further aggravated by varying street orientations. Some orientations are more advantageous than others for solar energy utilization. The following paragraphs discuss the ramifications of three differing street orientations.

1. **North-South Street Orientation.** Streets oriented north-south will generally result in north-south lot placement with the long axis of the lot in the east-west direction. Consequently houses will be sited in close proximity to one another along the north-south axis, causing shading and probable solar-access problems. In lower densities this may not be as much of a problem. The north-south orientation is good for primary streets, which may run perpendicular to residential streets.

2. **East-West Street Orientation.** Streets oriented east-west will generally result in east-west lot placement with the short dimension of the lot adjacent to the street right-of-way and the long axis of the lot in the north-south direction. If the buildable portion of the lot is wide enough, solar collection devices can be integrated easily into the south facade and roof with no appreciable shading. This is a good street orientation for streets adjacent to the majority of lots. The east-west street provides added solar-access protection.

3. **Diagonal Street Orientations.** Diagonal street orientations, especially those beyond 30° of the cardinal points, can cause a reduction in solar collection. A square relationship of the solar house to the lot may not be possible and may stress the overall parcel design. Depending upon the specific direction of the diagonal, either morning or afternoon shadows from adjacent houses may occur. Curvilinear streets may have the same limitations. A house can have a south orientation on a diagonally oriented plot depending upon the size of the lot.

Pedestrian walkways and bike paths also form an integral network within the neighborhood. It is possible to reduce energy and air pollution, which otherwise may occur because of the use of automobile transportation, by encouraging walking and bicycle riding. This can be accomplished with a well-thought-out nonautomobile movement system. In the example in Figure 6-6 some locations of the walkways and paths are parallel to the street grid and alleys and other locations are not. When they venture away from the automobile system, there is an opportunity for providing alternative routes that can link together many special neighborhood features.

Community-Service Considerations for Neighborhood Planning

Community services or facilities can greatly enhance a neighborhood. The extent to which these features are included within a neighborhood plan is dependent upon the size of the neighborhood and the economic base. It is very important in the planning of a new neighborhood to allow for the evolution of community features that initially may not be economically feasible. In time these community features may be realized. The type of community features will vary from neighborhood to neighborhood and will change over time as the community matures and grows. Some

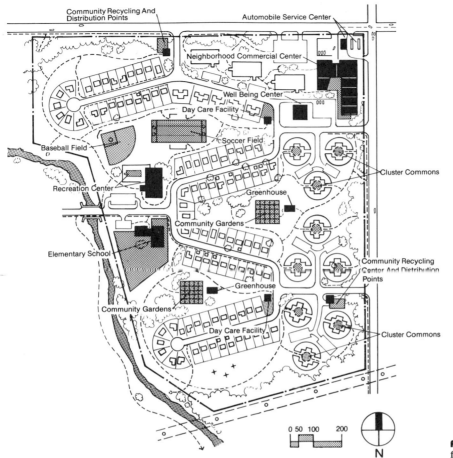

Community Recycling And
Distribution Points

Automobile Service Center

Neighborhood Commercial Center

Well Being Center

Day Care Facility

Baseball Field

Soccer Field

Cluster Commons

Recreation Center

Greenhouse

Community Gardens

Elementary School

Community Recycling
Center And Distribution
Points

Greenhouse

Community Gardens

Day Care Facility

Cluster Commons

0 50 100   200

N

Figure 6-7 Community
features.

features can be dispersed throughout the neighborhood, such as landscaping, but generally the features are more isolated. Figure 6-7 illustrates some community features for the example neighborhood plan.

Single community features of the plan are an elementary school located
near the center of the neighborhood adjacent to the stream, a community
space associated with the elementary school, community gardens located
adjacent to the stream and other open spaces, and a neighborhood recreation center and park located to the northwest. Community features distributed throughout the neighborhood include open spaces located in the
center of the subneighborhoods, two day-care locations, a number of
commercial facilities located to the northeast, and landscaping buffers
along the major streets. The commercial facilities could typically house
providers of goods and services, such as a small food store or cooperative,
bakery, well-being or medical center, beauty salon, hardware store, land-
scaping-supplies store, hobby shop, and coffee or herbal tea shop. Small
institutional or government buildings that may be appropriate at the
neighborhood level include a larger school, fire station, branch library,
and post office (especially convenient during the Christmas holiday).

Community gardens or allotments are gaining in popularity. Many people are interested in growing a variety of flowers, vegetables, and fruit- and nut-bearing trees in an area other than their personal garden or yard. In the areas designated for higher density, there may not be enough open space for individual plots and a community garden may be appropriate. The location of the community gardens in the example in Figure 6-7 was chosen for proximity to water, allocated open spaces, and users. Perhaps, a location near the housing clusters may encourage greater participation where private outdoor space is limited.

Solar-Access Considerations for Neighborhood Planning

The descriptive solar-access issue at the neighborhood level is providing context protection. Internal solar-access protection is extremely important but is better addressed at the subneighborhood or architectural levels. Context protection can be provided with any of the methods previously discussed in Chapter 3. They include bulk-plane, solar-envelope, and solar-fence zoning methods. Each method has its limitations. The bulk-plane method may be oversimplified and does not account for morning and afternoon shadows. The solar envelope can be complicated. And the solar fence is tied to property lines, which may have little relation to PUDs. For the purpose of the example neighborhood plan, a simplified solar-envelope zoning technique is used combining the bulk-plane, solar-fence, and solar-envelope concepts.

Several generic solar envelopes have been generated rather than a tailored solar envelope for each individual property. The envelopes vary depending upon the land use and density and their specific location within the neighborhood. The first zone protects the single-family detached residences and relates to the greatest number of properties. Within this zone of the neighborhood, the density is approximately five units per acre. Each property within the subneighborhood is 60 feet in the east-west direction and 140 feet in the north-south direction. With 25-foot setbacks in the front and rear yards and a 10-foot setback in the side yard, the buildable area is 40 feet by 80 feet. This aspect will encourage placement of the long axis of the house in either direction rather than limiting it to only one direction. This zone is basically a bulk-plane envelope defining the buildable volume of the site. The height limit at the southern line of the buildable area is 40 feet, and the height limit at the northern line is 20 feet.

The second solar zone is for the cluster housing located in the east and west of the neighborhood. Because of the higher density of fifteen units per acre, a larger solar envelope is required. The blocks for the clusters are approximately 4 acres and allow two clusters of a total of thirty housing units each. This solar zone is also a bulk-plane envelope defining a buildable volume with a southern height limit of 55 feet and a northern height limit of 25 feet. Streets to the east and west of the clusters further help in providing solar-access protection to the adjacent single-family residences.

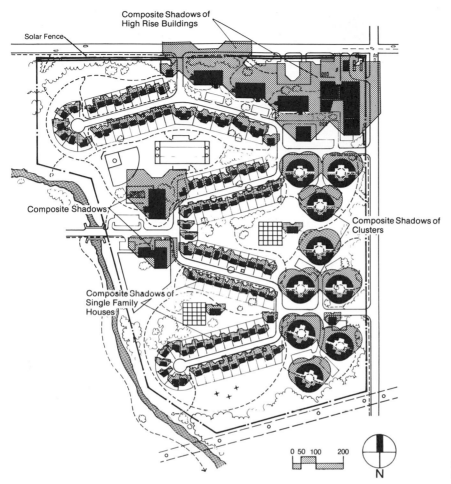

Figure 6-8 Neighborhood solar-access zoning.

The third solar zone is located near the northeast corner of the neighborhood and covers three small blocks near the center, where the elementary school is located. In this zone both commercial and higher-density residential features occur. The residential density is twenty-five units per acre. It is assumed that a major street is directly to the north of this zone and therefore can help protect residential development farther to the north should it occur. The solar zone used in this situation is a solar fence with a height limit of 25 feet on the north property line. There is no solar zoning needed for the other land features within the neighborhood as they do not allow for further development. Each of the solar-access zones are based upon shading on December 21 between 9 a.m. and 3 p.m. Figure 6-8 indicates the shadow patterns for each of the buildings and identifies solar zones within the neighborhood.

Utility and Energy Supply Considerations for Neighborhood Planning

The availability and capacity of various utilities are important factors in developing neighborhoods; however, the utility infrastructure generally

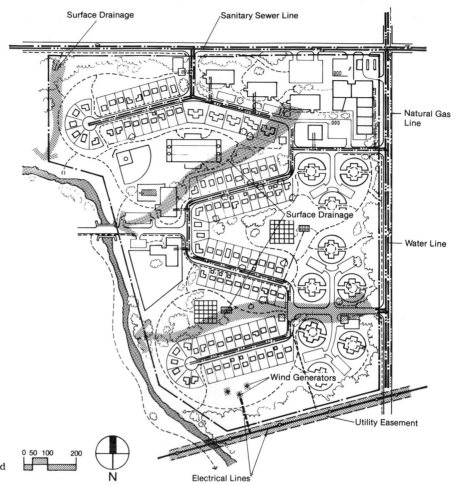

Surface Drainage

Sanitary Sewer Line

Natural Gas Line

Surface Drainage

Water Line

Wind Generators

Utility Easement

0 50 100 200

N

Electrical Lines

Figure 6-9 Neighborhood utility infrastructure.

has an indirect impact upon neighborhood form. In fact, all of the utilities—water, sanitary and storm sewers, gas and electricity, and telephone—are important in maintaining life support. Because these systems are so commonplace, they are quite often taken for granted. Simple, straightforward utility planning can save energy. Figure 6-9 illustrates the utility infrastructure for the neighborhood example.

There are three options presently available for neighborhood utility systems. They are (1) public systems, which generally belong to a larger utility network; (2) systems owned and operated by the community; and (3) private or individual on-site systems. If the new neighborhood is a part of a greater existing community, the public utility system will undoubtedly be the most inexpensive solution. In more remote areas, privately owned and operated utility and energy systems may be the only alternative. Community or perhaps neighborhood utility systems may in the long run offer advantages of the economy of scale while maintaining community control and using innovative recycling systems. This is especially true for water and sewage recycling.

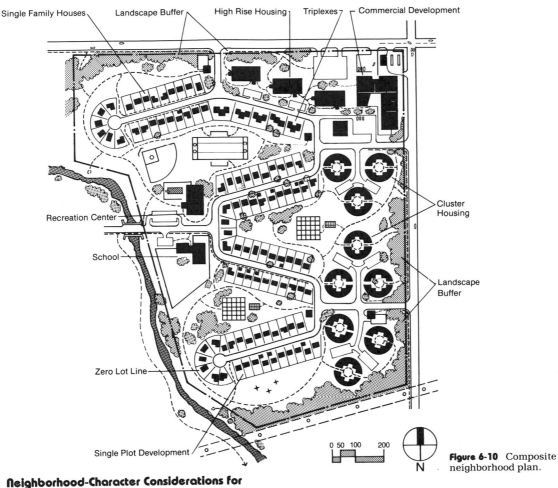

Single Family Houses · Landscape Buffer · High Rise Housing · Triplexes · Commercial Development

Recreation Center

School

Zero Lot Line

Single Plot Development

Cluster Housing

Landscape Buffer

0 50 100 200

N

Figure 6-10 Composite neighborhood plan.

Neighborhood-Character Considerations for Neighborhood Planning

Many factors can add to the overall character of a neighborhood, giving it a sense of place as well as aesthetic appeal. These factors may not directly relate to energy consumption but can affect the mobility of the neighborhood and the resulting energy use. The street, whether it is landscaped or used for parking; the utility lines, whether they are exposed or not; the styles of architecture, whether they are harmonious or not; and the structure or neighborhood organization, whether it supports a center and feeling of community or not—all contribute to the quality of the neighborhood character. And probably one of the most important factors relating to the neighborhood quality is the attitude of the residents. A positive attitude toward the environment will find expression in many ways. At the scale of the neighborhood, it can be extremely contagious.

Figures 6-10 and 6-11 illustrate a composite of each of the neighborhood elements: varied land use and density, east-west-oriented streets and bike paths, community facilities, planning for solar access, and a simple utility system. Landscaping, community gardens, off-street park-

Figure 6-11 Isometric.
(Illustration by Nate Kipnis.)

ing, a variety of housing types, and a strong boundary definition help form an identifiable neighborhood.

EHISTING NEIGHBORHOODS

In the year 2000, 80 percent of the building stock will be composed of buildings that are standing today. One of the most difficult challenges ahead is improving the energy efficiency of these existing structures. An even more difficult problem facing the residential sector is improving energy efficiency and strengthening neighborhood support. The move toward energy retrofitting existing residences in the United States in the last 10 years has been quite widespread. Nearly 20 million houses have been retrofitted to higher insulation levels with attic insulation, caulk-

Figure 6-12 Direct-gain and solar hot-water heating retrofit.

ing, and weather stripping. Many existing houses have been retrofitted with a variety of passive and active solar heating systems as well. Retrofitting entire neighborhoods, on the other hand, is a different proposition, involving more than the technical problems related to conservation techniques or solar energy systems. However, according to Christopher Flavin, ''In most countries it should be possible to bring energy-conscious design into the mainstream in 1990, and to reduce the fuel consumption of existing buildings by more than 30 percent.''[25]

Retrofits, Renovations, and Adaptive Reuse

Energy upgrading existing houses usually involves a variety of energy-conservation measures, solar domestic hot-water heating systems, and in many instances passive or active solar space-heating systems. No generalization about methods, other than energy conservation, can be made. Every house is unique and therefore requires a thorough investigation prior to identifying specific solutions. Often measures that can easily be incorporated with new construction are not economically feasible with energy retrofits. For example, providing additional wall insulation can be extremely expensive with existing older homes. Figures 6-12 through 6-15 illustrate energy-oriented retrofits to a variety of existing homes. In these examples energy-conservation measures have typically been employed, and solar hot-water heating, increased glass on the south, or sunspaces have been added.

Many communities have instigated winterization programs aimed at reducing energy consumption in existing buildings. Energy-conservation measures commonly associated with these winterization programs and general solar retrofits include increasing mechanical efficiency, reducing

Figure 6-13 Attached sunspace retrofit. (Courtesy of Phllip Tabb Architects.)

infiltration, and reducing winter heat loss and summer heat gain. Federal, state, and local grants or tax credits have provided added incentives to realize these goals. Below is a more detailed list of energy-conservation measures generally found to be cost-effective for existing buildings.[66] They do not require major construction or alterations, and they can be accomplished over several heating seasons.

1. Replace pilot light on gas furnace with electronic ignition device.
2. Add automatic dampers in furnace flue.
3. Replace existing thermostat with one with night setback.
4. Caulk, seal, and weather-strip all doors, windows, wall outlets, and other openings to the outside.
5. Provide additional roof or attic insulation if possible.
6. Install additional window glazings, storm windows, or movable insulation.
7. Add a solar domestic hot-water heating system with insulated storage tanks.
8. Upgrade fireplaces with insulated dampers, face glazings; or possibly replace them with more efficient wood-burning stoves.
9. Install whole-house fans for summer ventilation.
10. Provide summer shading devices, if they do not already exist, and energy-responsive landscaping.

Solar additions can differ from retrofits by providing more economical possibilities for incorporating greater energy conservation-measures and solar energy systems. It is more cost-effective to incorporate the energy measures along with the new construction. With solar additions, con-

Figure 6-14 Solar active space-heating retrofit.

Figure 6-15 Detached sunspace retrofit. (Courtesy of Phillip Tabb Architects.)

struction is not as inhibited by the existing structure; consequently, from the footings and foundations on up to the roof, energy-conservation steps can be taken throughout the structure. In the example illustrated in Figure 6-16, a single-family house located on the southern portion of the site has been renovated and retrofitted. In addition, a new attached housing unit has been constructed to the north in a north-south zoning arrange-

Figure 6-16 Solar retrofit and addition. (Courtesy of David Barrett, Boulder, Colorado.)

ment. On a sloped tower mounted on the hip roof is an active collector array for domestic hot-water heating. Both the new and the old blend to form an energy-efficient duplex. The project was designed by architect David Barrett.

One problem for energy retrofits or additions, aside from the technical problems, is associated with historic designations for either neighborhoods or individual buildings. Changes to these buildings need careful design and most often are required to undergo review from architectural review boards or special review committees with city planning commissions. This process can add more time and expense to a project. The additional time required may be as much as two months depending upon the approval process and schedule. Additional time for design and presentation and increased construction will raise the cost. Energy retrofits and energy-oriented designs need to be planned with sensitivity to existing neighborhood character.

Adaptive reuse is currently a popular method of development. Rather than tearing down existing buildings, new uses are envisioned for these buildings as they are renovated to meet contemporary needs. Some of these projects transform the older architectural style into a modern (or postmodern) image, while others have a strong respect for the original architecture. Both approaches can save money and energy. The greatest energy savings often come from the *embodied energy* associated with the original building in terms of the manufacturing of building materials, the transportation of materials to the site, and the construction process. A substantial investment of time, money, and energy are embodied within our present inventory of buildings—many of which are extremely suitable for renovation and reuse.

Existing Neighborhoods

People like old buildings and established neighborhoods. Maintaining and upgrading existing neighborhoods offer tremendous challenges. Many descriptive issues alongside the energy ones are important to understand in order to make the appropriate changes to existing neighborhoods. Existing neighborhoods, especially the more established ones, have evolved patterns that cannot be neglected, such as pedestrian versus automobile zones, character, and degree of transiency, to name a few. Neighborhood issues that may be affected with energy-related changes include zoning control, property value, traffic, utility service capacity, as well as character and image. The neighborhood as a whole needs to be considered with each building to avoid piecemeal changes. Changes to the existing neighborhood should be considered in relation to comprehensive community plans. Several planning strategies can contribute to a more cohesive transformation. In addition to a strategy of developing more energy-efficient buildings, they include (1) a strategy of density transfer, (2) a strategy for reducing automobile use, and (3) ways of encouraging shared or community spaces and activities.

The concept of density transfer for neighborhoods is simply the creation of areas of higher density along with greater open space. This is accomplished by building a variety of housing types within existing neighborhoods, upgrading existing structures, and in some instances demolishing unsuitable buildings in order to create more open space. This gives more variety in density and presents more choice in housing form. Densification, or increasing density, places greater demand upon existing utilities and services. However, it can reduce transportation energy consumption through reduced travel; the energy savings from this can be considerable, depending upon the location of new parcels of land.

Automobile use can be reduced first of all by locating neighborhoods closer to work areas, schools, and the larger community services. Automobile use can be reduced by providing more goods and services within neighborhoods so that residents do not need to travel as much to have access to these nonresidential uses. Automobile use can be further reduced by encouraging alternative modes of transportation, including walking, running, bicycling, minibusing, busing, and other means of mass transit. Further reductions in energy consumption of automobiles can occur with increased efficiency—a phenomenon that is finally being incorporated in both foreign and domestic automobile production.

Shared activities, facilities, and services, as mentioned earlier in this chapter, can contribute to reducing the duplication of energy consumption as well as land use. Mixed-use development should be encouraged as it can foster density transfer, shared facilities, and alternative transportation. Neighborhoods can be planned with diversity and vitality and be energy-responsive as well. With these ideas, static, single-use suburban development does not need to be considered the practical solution to our contemporary housing needs.

SUMMARY

It is quite evident that in the years to come the neighborhood will be seen as an extremely important planning scale. With the cycles of change and transformation, inner cities, suburbs, and rural communities alike can build strong, energy-efficient neighborhoods. This process can reduce energy consumption for both buildings and transportation. At the sub-neighborhood level, PUDs and city-block designs are limited more to energy savings for buildings. At the larger level of new and existing neighborhoods, both buildings and transportation can be addressed. The neighborhood can become more self-sufficient and more identifiable and can foster a greater sense of place.

RESIDENTIAL SETTLEMENT 7

As we become more concerned with land-use planning, it is important to recognize that natural, self-sustaining solar-powered ecosystems have a direct value to man for their life support and waste assimilation capacities as well as for their food, fiber, or recreational potential. EUGENE P. ODUM[46]

Solar energy is a dispersed source of energy; fuel is a concentrated source of energy. As a consequence, a dichotomy exists that strains the direction and growth of city form. Centralization versus decentralization is not a new problem; nonetheless, it is a terribly important one and one that needs to be considered in the broader context of the flow of resources and energy and the organization of residential settlements—both new and old. Existing center cities and supporting suburbs, new urban infill, and entirely new communities alike have opportunities for healthy change. This is an enormous task that cannot be accomplished overnight. Yet, vision and day-to-day evolutionary steps can transform the built environment into a more responsive place for habitation.

This is the fourth and last chapter in the sequence of discussion of increasing planning scales. Continuing the development of the last chapter, this chapter addresses building and transportation energy needs. However, this planning scale goes beyond the neighborhood scale to include groups of neighborhoods, more diverse community services, larger work and commercial environments, and larger institutional networks required to maintain community life. Divided into two parts, *solar planning concepts* and *holistic planning concepts,* the chapter describes applications of solar energy and other renewable resources and energy applications appropriate to larger-scale settlement.

SOLAR PLANNING CONCEPTS _____

Solar planning concepts for residential settlement are similar to those at the solar shelter, cluster, and neighborhood scales. Complications at the settlement level are the social, economic, political, and physical issues inextricable to planning for larger populations. As scale increases, basic planning elements tend to amplify. As a result, entire residential-settlement concepts are rarely entertained beyond the comprehensive land-use planning level. Physical form is presented on a project-by-project basis. For solar planning concepts to actually be used, they need to be integrated into the mainstream planning process, including the comprehensive land-use plans, subdivisions, and PUDs.

Presently, development trends are in the areas of urban redevelopment, urban and suburban infill, and new development in fringe areas. If these trends remain true for the next several decades, solar energy planning scenarios for all of them should be drawn.

Solar Concepts for Urban Redevelopment

Most urban areas in the United States were developed during the rise of the fossil fuel era, 1860 to 1960. Consequently, they bear the order and structure inherent to the fossil fuel economy—central energy systems for both communities and individual buildings, little concern for solar orientation, and large suburban developments linked by thousands of miles of streets, roads, highways, and expressways. In large urban areas with extensive city centers, great resources have been invested not only in creating the structure and form but in maintaining it. Application of broad solar concepts to these areas is not likely—at least in the near future. The energy needs of urban centers are quite different from those of residential neighborhoods. Transportation, lighting, cooling, and air-conditioning energy needs dominate the energy needs of buildings in the more dense urban centers. Heating, specifically solar heating, is often not that desirable.

Typically, the dense packing of high-rise buildings is not conducive to adequate access to the sun. Most high-rise buildings in winter are in shade in the morning or afternoon or both, making solar heating very difficult. During summer, almost all high-rise buildings need cooling and air-conditioning; and during winter, it is quite common for high-rise buildings to need both heating and cooling, which can usually be accomplished by relocating heat from the building core to the perimeter. For high-rise and many other nonresidential buildings, lighting and cooling make up approximately two-thirds of the total energy needs. Daylighting is being recognized as an extremely important strategy for larger buildings. By utilizing natural light, artificial lighting can be reduced dramatically. This in turn reduces the cooling load. According to a recent publication by the Solar Energy Research Institute (SERI), "Lighting accounts for about 20% of the total electrical consumption in the United States."[57]

Energy consumption in urban centers is different and requires a differ-

ent set of strategies. Mass transportation, mixed-use development, designing for natural daylighting, and energy-conservation measures appropriate to larger buildings dominate the energy picture at the highest densities. Providing solar-access protection to adjacent lower-density areas is also important. This is especially important when urban centers are adjacent to predominantly residential land. Shadows to the north-west, north, and northeast cast by large buildings can cause problems for solar-oriented residential development. Figure 7-1 is an aerial view of Boulder, Colorado, illustrating the organization, street patterns, and density.

Solar Concepts for Infill Development

Infill development is currently a popular alternative to both urban and fringe development. Infill projects are located throughout the city fabric,

Figure 7-1 Aerial photograph of urban center, City of Boulder, Colorado, 1979. (Photograph by ARIX.)

offering many choices in types and sizes of development. Infill developments between the more dense city centers and outer suburban edges lack some of the problems associated with the inner city and the outer city. However, the availability of urban and suburban land is diminishing, thus driving property costs higher and higher. It is difficult to generalize solar concepts for infill development. Although many remaining sites may be desirable, the majority of sites may present difficulties for the economic utilization of solar energy. Site size, shape, density, topography, and orientation can vary greatly, as can the uses of adjacent land. Consequently, solar energy concepts need to be considered on a project-by-project basis.

Infill projects are likely to be single-lot developments, PUDs, and block developments. Entirely planned infill communities are probably not possible because of the lack of adequate space. The larger-scale issues of infrastructure, movement systems, water, waste management, etc. will probably not be descriptive for infill development. Therefore, infill development is generally going to be piecemeal. Solar energy concepts can be applied to individual buildings and small groupings of buildings. Infill development may not afford the opportunity for large-scale change—if it is desirable. Areas in and around the central business district and off into the suburbs offer opportunities for infill development.

Solar Concepts for New Communities

Solar concepts for new communities offer exciting possibilities for new forms of living. Although many new communities may not exist in regions with favorable sunshine, those that do may be planned and organized for utilizing solar energy and implementing innovative concepts for shelter, transportation, and other community systems. The physical form of a new community can be adapted to the climate and weather patterns of the specific region for which it is planned. Sites for these communities are most likely to occur around the fringe of existing cities, forming a link between natural and rural environments and the existing urban environment.

Climate can play an important role in determining overall community form. In *Design With Climate,* Victor Olgyay presents climatic characteristics and conceptual approaches to community planning for four climatic zones in the United States. Although some of the concepts of Olgyay have recently been criticized as being too dogmatic, many of his ideas form a good beginning for the planning level. Similar to the climate concepts discussed in Chapter 4 for small-parcel developments, these community concepts are developed for temperate, cool, hot-arid, and hot-humid climate zones. In each scheme various densities of housing, recreational facilities, schools, administrative facilities, and commercial development are suggested. The location, shape, organization, open-space design, and landscaping are adapted for each climate zone. The sites are theoretical with a valley, waterway, and slopes. Figures 7-2 through 7-5, redrawn from the original Olgyay site plans, illustrate the community plans for these four zones.[48]

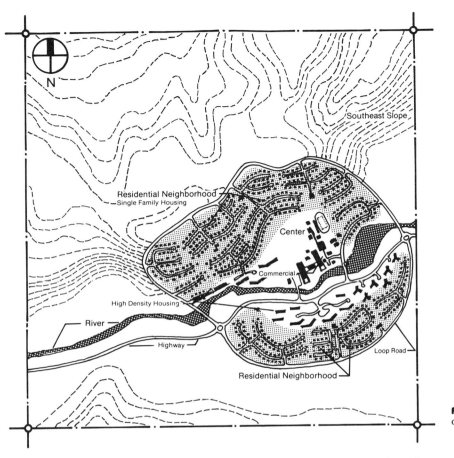

Figure 7-2 Temperate-zone community plan.

In the *temperate zone* the location of the community site is based on south and southeasterly orientation for winter solar radiation and summer breezes. The community is organized into a ring arrangement with public facilities in the center very much like the energy-oriented cluster development discussed in Chapter 5. On the north side of the housing neighborhood are tree belts blocking the winter winds. Major roads ring the residential neighborhoods and run in roughly the north-south direction. Secondary streets generally run in the east-west direction. Housing is typically in the form of single-family detached houses, row houses, or high-rise apartments.

In the *cool zone* the location of the community site is based upon a favorable exposure to solar radiation and protection from prevailing northwesterly winter winds. In this scheme, the community is organized in a more linear arrangement with most of the housing positioned on two southeast slopes. Public functions are located compactly between the housing sectors. Two large tree belts are positioned to the northwest of the housing sectors. Roads and streets in the cool-zone community are similar to those in the temperate zone. Housing is in the form of single-family detached houses, medium-density group housing, and high-rise apartments.

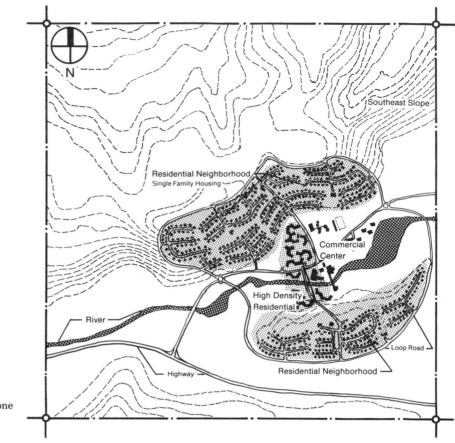

Southeast Slope

Residential Neighborhood
Single Family Housing

Commercial
Center

High Density
Residential

River

Highway

Loop Road

Residential Neighborhood

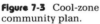

Figure 7-3 Cool-zone
community plan.

In the *hot-arid zone* the location of the community site is based on
proximity to the valley floor and water. The community is elongated
along the waterway and organized into several neighborhood clusters
with water and open space in the center of each cluster. The public facilities are organized similarly to the neighborhoods into compact courtyard
clusters. Landscaping is generally confined to individual buildings for
summer shading and the channeling of summer breezes. Automobile circulation is provided by major roads following the linearity of the overall
community and connecting the clusters. Streets go into the housing clusters. Housing is generally clustered into low- and medium-density residential areas with some high rises.

In the *hot-humid zone* the location of the community site is based on
proximity to undisturbed wind exposure and, therefore, is on the higher
plateau. The community is sectioned off into six residential neighborhoods. Landscaped open spaces, the higher-density housing, and the public facilities separate the neighborhoods. Major roads ring and pass
through the center of the community while streets form a network in both
the north-south and east-west directions. Housing is primarily in the single-family detached form with some medium- and high-density housing.

Labels on figure:
N

Southeast Slope

Residential Neighborhood

River

Loop Road

Commercial Center

Residential Neighborhood

High Density Residential

River

Residential Neighborhood
Single Family Housing

Neighborhoods Located on Valley Floor

Highway

Figure 7-4 Hot-arid-zone community plan.

Most of the housing is low-density and is spread out and elongated along the east-west axis. The public buildings are also spread out to allow the natural air currents to flow freely.

There are, in fact, more than four climate zones in the continental United States. These four solar community concepts are simply generalized planning models. In the next several decades, the type of new community most likely to occur is the satellite community. These new communities may be bedroom settlements for larger urban centers, new settlements organized around new industry, or expansions to the many small towns. And they are likely to be disposed throughout the United States. They afford the opportunity to be planned for solar energy and the natural environment. This kind of community is aptly envisioned in Michael Corbett's book entitled *A Better Place to Live.*

> The pattern I think we should work toward is one of small, relatively moderate-density towns with enough distance between them to give lower density overall. Moderate density within the town would have the advantage of providing stimulating social contact and eliminating most of the need for automobiles. Low regional density would reduce air pollution, allow local agriculture production for each town's needs, permit

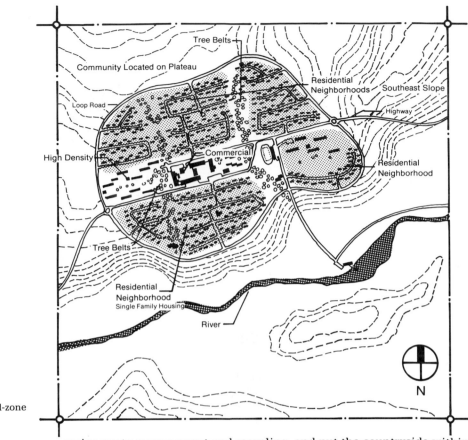

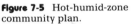

Figure 7-5 Hot-humid-zone community plan.

easier waste management and recycling, and put the countryside within easy walking distance of every home. This is essential if we are going to live within the limits of renewable energy supplies, maintain a healthy environment and ensure sustainable food production.*

The student-designed new community that is illustrated in Figures 7-6 through 7-10 incorporates many of the planning and design ideas discussed in the last four chapters. Planned for nearly 9000 residents with approximately 3200 housing units, the single-plot developments, clusters, high-density housing, commercial areas, institutional facilities, recreation offices, and warehouses have all been designed to form a self-sufficient community. The location of the community is presumed to be in a temperate climate zone between latitudes 35 and 45° north. A more detailed listing of the community facilities and features follows:

1. Medium-income single-family detached housing (1080 units)
2. Zero-lot-line detached housing (110 units)
3. High-income detached single-plot development (60 units)

*Reprinted from *A Better Place to Live* © 1982 by Michael N. Corbett.[15] Permission granted by Rodale Press, Inc., Emmaus, Pa. 18049.

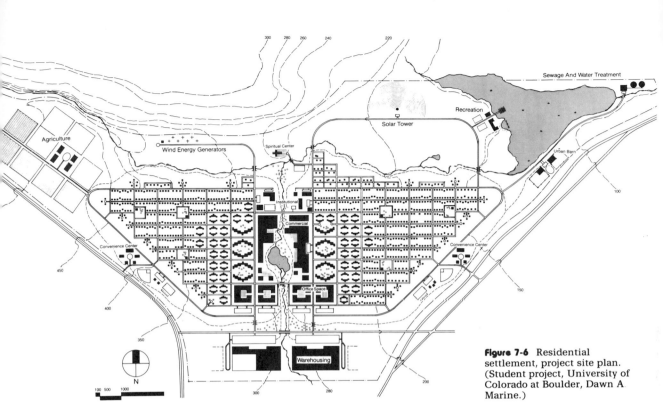

Figure 7-6 Residential settlement, project site plan. (Student project, University of Colorado at Boulder, Dawn A. Marine.)

4. Low-, medium-, and high-income cluster housing (1170 units)
5. High-rise housing (750 units)
6. Commercial center
7. Convenience centers (2)
8. Professional office blocks
9. Warehouses for community goods (750,000 square feet)
10. Resource-recycling center and distribution points
11. Neighborhood centers, schools, and day-care facilities
12. Recreational facilities and recreational open space
13. Institutional facilities (post office, fire station, library, financial center, health-care center, and community-maintenance center
14. Nondenominational spiritual center
15. Community greenhouses and growing fields (70 acres)
16. Community farm (20 acres)
17. Community solar power-generating system (5 megawatts)
18. Community waste-treatment system
19. Pedestrian walks and bike paths
20. Minibus transportation system

   The site for this project is located to the north of a moderate-sized city and is bounded by major highways on the east and west, an access road to the south, a small lake to the northeast, and a waterway and a mesa to the north. The site slopes up and down from the west to the east. The

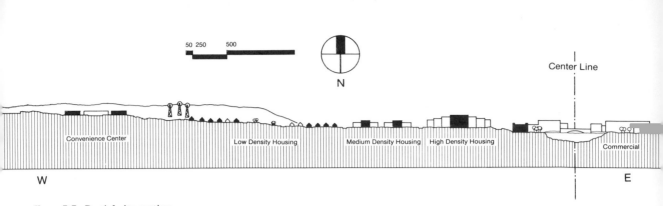

50 250   500

N

Convenience Center   Low Density Housing   Medium Density Housing | High Density Housing   Commercial

W   E

Center Line

Figure 7-7 Partial site section.

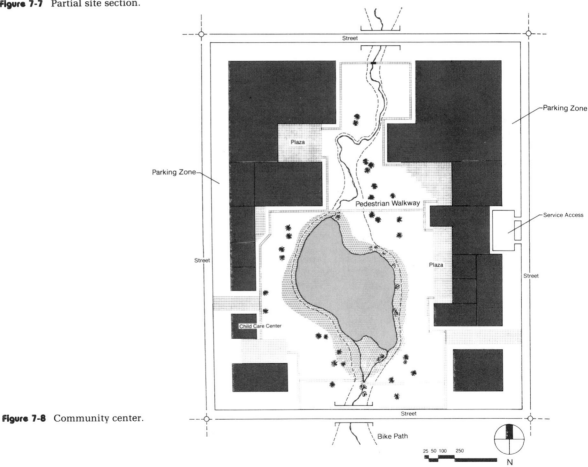

Street

Parking Zone

Plaza

Parking Zone

Pedestrian Walkway

Service Access

Street

Plaza

Street

Child Care Center

Street

Bike Path

25 50 100   250

N

Figure 7-8 Community center.

total gross area of the site that the community occupies is approximately 780 acres for a gross density of 4.1 units per acre. This includes open space surrounding the community, streets, neighborhood parks, and commercial center. Elongated on the east-west axis, the site has excellent solar access.

The central commercial district separates the community into two

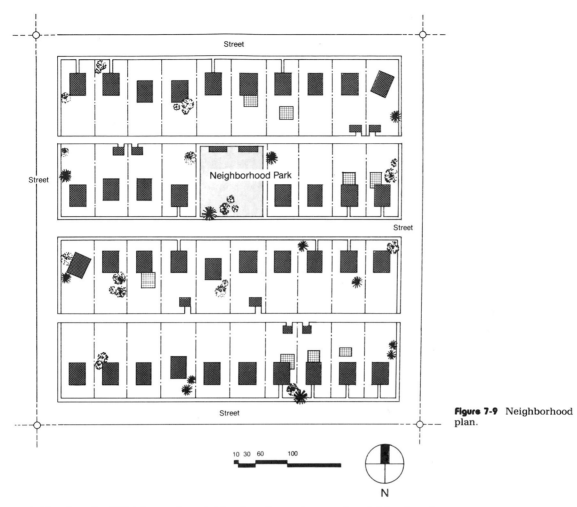

Street

Street

Neighborhood Park

Street

Street

Street

10 30 60   100

N

Figure 7-9 Neighborhood plan.

neighborhoods. Within each neighborhood are three subneighborhoods. Density decreases from the center outward. There are four major automobile entries, a loop road, major streets oriented in the east-west direction, and minor streets oriented in the north-south direction. The site is slightly hilly, which softens the effect of the rectangular street grid. Refer to the partial section in Figure 7-7. Subneighborhoods center around open spaces with elementary schools and day-care facilities. The office blocks are located to the south of the commercial center and are planned in a courtyard configuration for maximum daylighting. The warehouses and recycling center are located across the major east-west road to the south. A recreation center and the community farm are located adjacent to the lake.

The solar energy features of this planning scheme include organization and layout for complete solar access for all the community's buildings and a solar power-generating system. The plan uses building designs previously discussed in the book, that is, single-plot developments, cluster

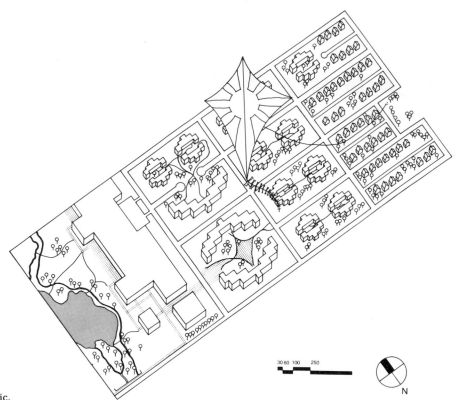

Figure 7-10 Partial isometric.

housing, and high-rise housing. Solar-access protection follows a hierarchy of scale with lower densities having greater protection.

The solar power system is based upon the solar power-tower concepts developed by McDonnell Douglas, Martin Marietta Aerospace, and Honeywell. The concept involves a vertical tower, which holds a boiler through which a working fluid passes. The working fluid, usually water, is transformed into steam and thereby produces electricity. A large field of tracking mirrors is positioned around the tower in order to reflect sunlight to the boiler during most of the daylight hours. The solar radiation is reflected and concentrated to create higher temperatures—as much as 1000°F.

The largest solar electric plant was completed in Daggett, California, producing 10 megawatts of electricity for some 5000 homes. On the desert floor are arrays of mirrors, 1818 in all, focused upon a 300-foot-high tower. A similar scheme was built earlier in a small town in south-central France (Odeillo, France) in 1971. It features a south-facing field of heliostats that automatically track the sun and focus solar radiation to a ten-story north-facing parabolic dish of mirrors (a receiver) that is integrated into the building form. The parabolic dish concentrates the solar energy further and reflects it to a small boiler in a separate structure as pictured in Figure 7-11. Although the solar furnace in Odeillo does not

Figure 7-11 Solar furnace, Odeillo, France. (Photograph by Phillip Tabb.)

operate a turbine, it does provide a great deal of information about high-temperature applications of solar energy.

Each of these solar power-generating schemes is based upon the same system: the field of mirrors or heliostats, the boiler, and the conventional steam turbine-generator. Larger solar power towers than the one in Daggett, California, are planned for the future and will produce as much as 50 or even 100 megawatts of electricity. Power towers rely on direct solar radiation and, therefore, are limited to locations with a high percentage of sunshine. There is also a practical limit to the size of a system. For example, according to Jan F. Kreider, author of *Medium- and High-Temperature Solar Processes,* large land areas are required for solar collection. In fact, a square mile would be required for a 100-megawatt power plant.[33] Should the solar power tower or a similar solar electric system become more economically feasible, it is most likely to be planned with smaller communities in favorable locations with available land. For the student example previously mentioned, a 5-megawatt solar power tower has been suggested to provide energy for the backup heating needs as well as the many other uses of electricity.

A third scheme for producing electricity from solar energy is the use of photovoltaics. Photovoltaics, simply put, are usually made of silicon crystals bonded to boron and phosphorous to form a small cell. When sunlight strikes the cell, a small current of electricity is generated. The photovoltaic cells are linked together within a collector and can either charge a battery storage array or be used directly. To some, this scheme offers one of the greatest potentials because of the more direct conversion of solar radiant energy to conventional electric energy. Additionally there are no moving parts, which reduces maintenance and increases longevity. Previous schemes required the conventional steam turbine-generator to convert thermal energy into electricity. This conversion process has its inefficiencies. According to Christopher Flavin, approximately 10,000 residences presently have some use of photovoltaics.[26]

The primary drawback to photovoltaics has been cost. Even though the cost has been steadily decreasing over the last decade, it is prohibitive except for very expensive homes or those in remote areas. The high cost is due to expensive materials, an energy-intensive process of production, and, to a certain extent, the additional parts and equipment necessary to integrate the entire system to a conventional house (array structure, wiring, converters, battery storage, etc.). Presently, the average cost of photovoltaics is approximately $10 per peak watt. To provide electrical energy for a typical energy-conserving house, between 100 and 175 square feet of photovoltaics are required at current efficiencies of approximately 10 percent. Using high-technology manufacturing techniques, millions of photovoltaic cells can be produced within the very near future. With solar utilization for single buildings along with the potential for solar electricity generation, a great amount of fossil fuel can be conserved. However, the use of photovoltaics is presently limited to areas where conventional energy sources are impractical or unavailable.

Putting all the complicated pieces together to form a community is not an easy task. The capital needed to plan and build a community is enormous. Compared to all of the complex issues of town planning, the solar energy considerations may take a secondary role. This could be a big mistake. Solar energy planning concepts need to be considered from the beginning. They need to influence city form and city planning processes. A simple lesson to be learned is to design appropriate units of shelter, that is, units with an appropriate response to the human needs, site, local materials, and construction techniques. For example, in western France along the Atlantic coast is an area called Bretagne (Brittany), where the basic style of architecture has been observed for hundreds of years. It is simple and direct. Rectangular plans with a gable roof and dormers with operable shutters form the primary design. Most of the homes are sun-

Figure 7-12 Bretagne country house.

Figure 7-13 Bretagne subdivision.

tempered or passive-solar. The massive perimeter walls are made of stone, brick, or concrete block with stucco. Refer to Figure 7-12. These homes stand handsomely alone in the country or along the ocean, or they relate well in groups as pictured in Figure 7-13.

The key is simple, straightforward design, construction method, and style. The primary forms are very similar. Variety and individuality are to be found, but, from individual shelter to community, there is a continuity of style that responds to the specific climate of the region. American housing, on the other hand, has been too occupied with "individual expression," and a grouping of many homes usually lacks homogeneity. The relationship between the single home and the context of homes should be a great challenge to planners and designers of residential settlements.

HOLISTIC PLANNING CONCEPTS

The distinction between solar and holistic planning concepts is that holistic planning concepts include solar energy utilization in conjunction with other renewable energy and resource conservation approaches. Holistic planning is broader, more encompassing. Ekistics, as studied by C. A. Doxiadis, identifies the major elements of holistic planning: nature, the individual human, society, shells, and networks. In more practical terms, the ekistic elements of residential settlement may include open spaces, food production, transportation, energy-responsive shelter design, and landscaping. Human needs and aspirations are considered in the planning process, and the social sciences, shelter design, architecture, and urban design are taken into account. What is important with holistic planning is the integral relationships among all of these elements as they form a cohesive whole. Two elements of the ekistics model, nature and networks, are heavily influenced by energy planning concepts.

Food Production

One of the more interesting offshoots of solar planning is food production, especially suburban and urban food production. Since the industrial revolution, cities have become more segregated from food production. As long as large agricultural businesses have the resources and energy, great quantities of food can be produced. However, today, most agribusinesses are highly dependent upon fossil fuels and chemical pesticides for survival. Therefore, a rise in fuel costs will surely affect food costs. To offset this potentially dangerous situation, many individuals have sought alternatives.

The idea that human systems can integrate with and contribute to biological systems rather than damage them was investigated by the "new alchemists" in the early 1970s. The brainchild of John Todd, the New Alchemy Institute tried a new concept for a biotechnical system for supplying food and energy from a compact, self-contained environment. The first experiment occurred at Woods Hole, Massachusetts, where the integration of a combination greenhouse, fish farm, and renewable energy system was undertaken. The objectives of the experiment were (1) to create small-scale, decentralized agricultural systems as opposed to large agribusinesses; (2) to respond to ecological rather than economic considerations; (3) to use simple technologies that were comprehensible; (4) to bring the technologies to all users, including the poorer populations; and

(5) to produce reliable, rather than record, yields in the greenhouse and fish ponds. Figure 7-14 illustrates one of the greenhouses, the aquaculture unit, and the wind generator at the Woods Hole farm.

A second experiment took place on Prince Edward Island in the Gulf of St. Lawrence in Canada. The building of the project, or the Ark, as it was called, integrated the greenhouse, the fish farm, and the passive solar energy systems with housing in a three-bedroom house. The ecological design was based upon natural ecosystems in which there was an interdependency between fish communities and supporting plant life. The projects at Woods Hole and Prince Edward Island relied on solar energy and wind power. Both experiments were small and did not address the magnitude of problems associated with the large-scale production of food. However, the experiments of the new alchemists helped open up a more ecological method of food production. It will be interesting to see in the years to come the changes in agricultural methods as fossil fuels become more and more expensive.

There have been many efforts by individuals to produce their own food. Many private gardens and greenhouses have been producing a great deal of food throughout the United States. A 1979 Gallup poll showed that 33 million American households produced some of their own food. Garden plots vary in size from several hundred square feet to several thousand square feet. According to William Kaysing, author of *First-Time Farmer's Guide*, the size of a garden (or greenhouse) is arbitrary; however, a garden plot of 1000 square feet in most areas could provide plenty of food, primarily vegetables, for a family of four.[31] Growing seasons vary from region to region, thus limiting the amount of production. Typical of home-grown vegetables are loose-leaf lettuce, peas, spinach, parsley, rad-

Figure 7-15 Passive solar greenhouse. (Courtesy of Phillip Tabb Architects.)

ishes, carrots, beets, cauliflower, cabbage, pole and wax beans, corn, head lettuce, onions, and tomatoes. Combined with fruit- and nut-bearing trees, a home garden can produce a large amount of food.

Attached or detached greenhouses can provide solar heating as well as food production year-round. In addition to vegetables, the solar green-house can house seedlings, bulbs, flowers and other bedding plants, herbs as well as house plants. Solar greenhouses can vary in size depending upon the desired amount of produce. Cost is another important factor in determining the greenhouse size. If the greenhouse is too large, it may cause overheating. Certainly, the greenhouse size is going to be considerably smaller than a family garden. Figure 7-15 is an interior photograph of an attached solar greenhouse of approximately 120 square feet. Stepping planter beds, both inside and outside, connect the level of the back yard with a basement-level greenhouse 6 feet below grade. This partially earth-sheltered greenhouse takes advantage of heat stratification, with more constant temperatures in the bottom third of the greenhouse, where most of the plants are, and much higher temperatures at the top, which acts strictly as a solar collector.

Commercial and communal greenhouses can offer a tremendous value to a neighborhood or community. They too can vary a great deal in size—from 1000 square feet to several acres. The commercial greenhouse illustrated in Figure 7-16 is approximately 2 acres. The greenhouse primarily produces chrysanthemums for Great Britain. The total operation of nearly 10 acres of greenhouses in Titchfield, Hampshire, in southern England is one of the largest operations producing chrysanthemums in the world. Communal or community greenhouses can function effectively in the more urban areas. Care and maintenance can be shared by a large popu-

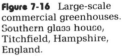

Figure 7-16 Large-scale commercial greenhouses. Southern glass house, Titchfield, Hampshire, England.

Figure 7-17 Community gardens in Holland.

lation. Community greenhouses can be associated with community gardens or allotments as well. Many of the energy savings gained by these greenhouses and gardens are hidden and difficult to quantify, but they include savings in production, transportation, storage, and distribution of food and other plant life.

The community gardens pictured in Figure 7-17 are typical of the allotments found in Europe. They usually occur on public land along greenbelts, flood plains, or open spaces at the outer edges. The community gardens are divided into plots, most often rectangular, and placed in rows.

Often garden sheds are provided for storage of tools and planting materials. Many families rely on the food produced by these gardens as the cost of fruits and vegetables is always on the increase. Without an allotment, many families would go without these foods.

Alternative Transportation

Transportation systems have been a major force in city planning especially since the turn of the century. Railroads, automobiles, airplanes, taxis, buses, and other transportation systems have literally structured city form and have become an extremely visible part of the urban environment. Many of the newer developed cities in the United States, such as Los Angeles, Dallas and Fort Worth, Houston, and Denver, have evolved simultaneously with the dramatic growth of the automobile industry. In fact, as much as one-third of the city land in many of these cities is given up to the automobile and its supports—roads, parking, service stations, dealerships, etc. The petroleum shortages experienced in the early 1970s were cause for investigation of alternative transportation means or at least improved forms.

The most visible changes inspired by energy conservation occurred in automobile size. Largely because of the influx of smaller foreign cars, American automobile users slowly replaced large overpowered cars with compact, subcompact, and small economy cars. Gas mileage increased from the teens to as much as 40 or more miles to the gallon for out-of-city driving. In the mid-1970s a 55-mile-per-hour speed limit was imposed on all highway travel. For several years this limit was respected by nearly all drivers to reduce fuel consumption. However, today many of the more anxious drivers have stepped up their speed. Most travel occurs within a very short distance from home—approximately 7 miles. As a result, the highway speed limit of 55 miles per hour does not have as far-reaching effects as previously expected. However, a reduction in the number of automobile accidents is a welcome side effect.

The profusion of automobiles has caused severe problems for the urban environment. Safety seems to be the most serious problem—for drivers and passengers as well as pedestrians. Conflicts between pedestrians and drivers have been serious for a long time. Partially prompted by the energy shortages and the visibility of outdoor air pollution, planning for pedestrians is a much more important consideration today. Solutions to the energy and transportation issue tend to follow similar lines throughout the developed world. Four overall strategies seem to dominate: (1) smaller, more-efficient automobiles and alternative fuels; (2) more-effective mass-transportation systems; (3) mixed-use development; and (4) planning for pedestrians. These strategies in concert with one another should certainly relax the problem.

Since much of the industrialized world enjoys the automobile for personal transport, the advent of the electric vehicle seems to have promise. Introduced several decades ago, the electric car became more visible after the Arab oil embargo. Dozens of companies sprang up throughout the United States. The electric cars were small, typically around 5 feet wide

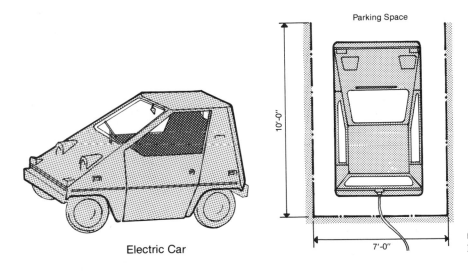

Parking Space

10'-0"

7'-0"

Electric Car

Figure 7-18 Electric car and parking place.

by 8 feet long. Refer to Figure 7-18, which illustrates an electric car and its parking place. Note that the parking place is virtually half the size (7 feet by 10 feet) of the parking place for a conventional vehicle. Several problems have plagued the electric car. The first is its range, or travel distance per electric charge. Typically, the range is between 35 and 65 miles. The second problem has been cruise speed, which ranges between 35 and 70 miles per hour. Most electric cars actually cruise at the lower speeds. The third problem has been the battery-recharging time. Depending upon the type of electrical service, 15 to 30 amperes, the time required to charge an electric car is between 6 and 12 hours. Batteries are bulky, heavy, and costly, and they have to be maintained.

Despite the limitations of electric vehicles, they do offer some advantages. They are quiet and extremely energy-efficient. They can operate at a cost of 2 cents per mile. Their purchase price is between $6000 and $18,000 (1983 dollars), which is similar to that of conventional cars, but most electric cars do not have the same quality of engineering and comfort found in conventional cars. The smaller electric vehicles usually are priced toward the lower end of this cost range. The electric vans and larger cars tend to be more costly. At any rate, they seem to be a perfect replacement for the larger, energy-guzzling, internal-combustion automobiles. Yet their popularity has not caught on. Some speculate that this is due to the relatively low cost of gasoline and oil and the reluctance of the large automobile industry to change. Nevertheless, the future probably holds greater use of the electric car—most likely in the first quarter of the twenty-first century.

Other alternative fuels for transportation at the present are not practical and lack any market penetration. Ethanol and methanol (wood alcohol) are agriculture-intensive and therefore expensive. Synthetic fuels (synfuels) are still very expensive. Hydrogen fuels for common transportation are inefficient and need to be cooled, which adds greater complexity to the system. However, each of these alternatives will be further

Figure 7-19 Boulder mall.

researched and demonstrated even as gasoline remains the preferred fuel for the time being.

Greater attention to the pedestrian environment may help reduce automobile travel and subsequent energy consumption. Designing streets for people is not a new concept. The Agora in Athens, for example, was designed to encourage social interaction. Travel on land, then, was largely accomplished by foot. Modern advocates, like Jane Jacobs, have been alerting us to the virtues of the street for decades. The street once again is being seen as a social setting. Figure 7-19 is a photograph of the pedestrian mall in downtown Boulder, Colorado. The project was com-

Figure 7-20 Pedestrian environment.

pleted in the mid-1970s. The architecture is credited to Everett, Zeigel, Tumpes & Hand, the urban design and landscape architecture to Sasaki Associates, and the graphics and industrial design to Communication Arts.

The Boulder mall is four blocks in length. It replaced a two-way east-west street. Automobiles are not permitted on the mall except at cross streets, where the automobiles must travel slowly and yield to pedestrians. Although bicycle riding is not permitted on the mall, there are many forms of alternative transportation: walking and running, skateboarding, riding in baby buggies, rollerskating, and sledding in the winter. During the summer months, a horse-drawn carriage travels around the mall. Figure 7-20 is a photograph of a children's play area near the center of the mall.

In the push for space within public networks, the automobile has dominated. Bicyclists and pedestrians have not had adequate response to their needs. As a general rule, the higher the speed of the system, the greater the space needs. Volume of traffic also influences the area requirements for these networks. Because bicycles and pedestrians travel at slower speeds and are still at lower numbers than automobiles, they receive less space and attention. Planning and transportation decisions made today will affect the future. If we are to evolve to alternative means of transportation, clearer direction is needed in order to influence current processes.

TABLE 7-1
Present Resource Recycling

Original Product	Recycled Product
Computer paper	High-grade paper, such as toweling, toilet paper, etc.
Mixed paper	Tar paper, cereal boxes, etc.
Newspaper	Newspaper (can be recycled approximately ten times)
Glass	Bottles (primarily beer bottles)
Aluminum cans	Aluminum cans or sheets
Motor oil	Industrial-grade motor oil or building heating fuel
Steel cans	Low-grade steel
Plastic	Low-grade petrochemicals
Cardboard	Cardboard and paper sacks
Building materials	Building materials
Automobiles	Scrap metal and auto parts

Resource Recycling

Waste from industrialized nations is tremendous. In the United States alone, approximately 4.3 billion tons of solid waste is produced each year. This amounts to nearly 20 tons per person each year. Waste recycling is not new. During the two world wars, many individuals, families, and groups saved and recycled many products in order to help in the war efforts. Today, there is not as much enthusiasm for recycling; however, some states and communities have rekindled the recycling spirit. California, Oregon, and New Jersey have taken the lead in resource recycling.

The kinds of materials suitable for recycling are varied, and the list is growing as new processes and reuses are created each day. Typically, various grades of paper, glass, and aluminum cans make up the primary recyclable items. Plastic, scrap metal, motor oil, and steel cans are becoming more economical to recycle as well. Table 7-1 identifies the materials that presently are being recycled.

Several factors are involved in realizing a successful recycling process: (1) public involvement; (2) adequate equipment, including baling systems, shredders, and trucks; (3) adequate staffing; and (4) healthy markets (buyers of recycled material). Recycling processes are more likely to survive at the community, city, and/or county levels, especially if they can tie into existing waste-removal and landfill systems. Recycling can certainly save a great deal of energy. It is difficult to determine the full impact of recycling, but the years to come will see more and more state and community involvement. The ultimate success, of course, is dependent upon the ability of private consumers to maintain recycling in the home and at work, the ability of the public to organize and support recycling systems or companies, and the ability of industry to continue to use and to create new uses for recycled materials.

Holistic Planning Checklist

A checklist is presented here in the hope of providing a planning guide. It is broad and, of course, does not encompass all of the planning needs (especially the economic and political needs) for a process as complicated and time-consuming as that demanded by the scale of community design. Nonetheless, this checklist emphasizes energy-conservation, solar energy, transportation, and integrated systems that may be applied to the planning of a solar community.

1. Energy-conservation techniques
    a. Energy conservation for single buildings
        (1) Reducing inefficiencies of conventional systems
        (2) Reducing energy demand
        (3) Reducing energy load
    b. Reducing inefficiencies in utility systems
        (1) High-grade energy production
        (2) Distribution
2. Solar energy considerations
    a. Planning for solar access
        (1) Street orientation
        (2) Solar-zoning concepts
        (3) Lot configuration
    b. Utilizing solar energy for single buildings
        (1) Incremental passive space-heating systems
        (2) Domestic hot-water heating systems
    c. Utilizing solar energy for electricity
        (1) Solar turbine systems
        (2) Photovoltaics
3. Transportation considerations
    a. Improving efficiency of transportation vehicles
        (1) Improving internal-combustion automobiles
        (2) Commercializing electric vehicles
    b. Improving mass transportation
    c. Increasing pedestrian planning
    d. Encouraging carpooling
4. Integrated planning techniques
    a. Using energy-responsive land-use planning
        (1) Densification
        (2) Mixed-use development
        (3) Energy-responsive landscaping techniques
    b. Integrating agriculture with communities
        (1) Reserving some fringe open space for agriculture
        (2) Encouraging allotments within the inner city
    c. Providing resource recycling
        (1) Community networks and systems
        (2) Providing storage space and efficient processes
        (3) Providing technologies in kitchen service centers
        (4) Providing convenient pickup methods

Figure 7-21 Romantic city of the past, Portmeirion, Wales.

Figure 7-22 Solar city of the future. (Robert McCall, artist.)

CONCLUSION

By looking at the changes that have occurred over the last several hundred years, we may be able to understand the magnitude of change that is possible ahead. The appropriate planning steps may then be more apparent. Less than 500 years ago Columbus set out to find a western route to Asia; soldiers fought with musket and cannon; and travel was accomplished by boat and horse. Today we have spacecraft traveling out of our own solar system. Wars are fought with the help of satellite communication systems. Many people can travel by automobile, airplane, or mass transportation. People can use home computers for entertainment, work, and education. The next several hundred years could bring incred-

ible changes in technology, deeper understanding of our universe, as well as dramatic physical changes to the built environment right here on earth. There may be healthy communities living in outer space. We may be traveling to faraway solar systems. The earth could be replanted with new forests and lush gardens, and our residential settlements could be completely revitalized.

However, we are very much a "now" generation; that is, we are very much involved with the present. Progress appears slow, and the conservative routines of everyday life appear to predominate over large-scale visions of the future. Even though we are curious about the future and excited by the science fiction put forth by the media, we are ambivalent toward rapid change. We live in a period of time with the greatest technological development; but many of us question the role of technology in the solving of our problems. The net effect of much of our technology is more damaging than beneficial. This is especially true for many of the military and energy technologies we have chosen to develop. The role of solar energy is still unclear. It has not proved to be the panacea it was initially thought to be. Perhaps this is due to a rather clumsy start.

The difference between the past and the future scenarios seems vast. The image in Figure 7-21 is that of Portmeirion, Wales. It is a romantic city with suggestions of Italian architecture of centuries ago, and it is emotionally compelling. Portmeirion, which was built in this century, is the work of architect Clough Williams-Ellis, and it is virtually a copy of Italian architecture created centuries before. In contrast, Figure 7-22 is a painting by Robert McCall. It portrays a visionary solar city of the future. It is stimulating and exciting with suggestions of high technology and a connection with outer space. Solar energy design has a place in the evolution of our cities both new and old. Differences of climate and culture will affect the designs, but there will always be a certain timeless quality.

We possess the ability to visualize our future while nurturing that which we have already created. For the decades immediately ahead, our choices of energy development will have profound effects upon our residential settlements. The decisions surrounding these choices should be cause for great concern, for they, most certainly, will set the tone for things to come. We have been blessed with the opportunity to visualize our earth—the whole earth. Whether or not to be fully creative with this gift is the choice that lies before us.

SOLAR-ACCESS CALCULATIONS

CALCULATION PROCEDURES

Under certain circumstances, designing for solar access can be intuitive and merely calls for common sense. This may be true for the conceptual design phases for a straightforward project. However, as a project gains complexity and greater detail, more precise figures may be required as the visualizing process becomes too vague. For example, site conditions may be difficult, densities may be high, program requirements may be demanding, or the architectural form may be complicated. As a result, procedures for determining shadows and solar-access protection need to be undertaken. Exact dimensions need to be determined.

In this appendix simple calculation procedures are presented for a variety of problems associated with solar access. Five separate problems have been identified, and specific procedures have been generated for determining (1) south-facing window overhangs, (2) sawtooth active or passive collector spacing, (3) building shadow lengths and masks, (4) building shadow lengths on sloping sites, and (5) building shadow lengths on sloping sites with off-axis slopes. Accompanying each procedure is an example problem. Tables with information necessary for the calculations can be found at the end of the appendix.

Overhangs

The purpose of an overhang will vary from climate to climate. In hot-humid climates, an overhang may be needed to provide shading all year long. In a hot-arid climate, it may be needed to provide shading in summer, spring, and autumn. In temperate and cool climates an overhang may be needed to provide shading only in the summer months. The shade protection that an overhang provides is determined seasonally and hourly. In other words, the amount of protection is a function of the months of the year and the times of the day that mark the change from overheating to underheating or vice versa. Early morning and late afternoon sun are relatively weak compared to noonday sun. As a climate becomes hotter, the daily duration of shading becomes greater.

The first step in determining the size of an overhang is identifying the duration of shading both seasonally and hourly. The seasonal period can be found by determining the monthly heating and cooling degree-day crossover. The hourly period can be found by analyzing hourly data if it is available. Overhangs are not as effective in spring and autumn because daily temperature averages may fall well below the comfort range while midday temperatures may be extremely high. Solar altitude and azimuth angles can be chosen from the time of day and the day of the year.

The next step is determining the form of the overhang. Fixed horizontal overhangs are excellent for providing shading during the midday hours. However, as the position of the sun moves lower and farther east in the morning and west in the afternoon, the horizontal overhang becomes ineffective. Therefore, vertical overhangs may be required. Figure A-1 illustrates the design of an overhang for a south-facing window located at 40° north latitude. The equation for a horizontal overhang is as follows:

$$D = H \cot Al \qquad \text{or} \qquad D = \frac{H}{\tan Al}$$

where D = distance of overhang
$\quad\ H$ = height from bottom of glass
$\quad\ Al$ = noon altitude angle

If the overhang is one that slopes upward, rather than a horizontal one, the calculation is a little more complicated. As illustrated in Figure A-2, use the following equation:

$$D = \frac{H \cos Al}{\sin C} \cos B$$

where D = horizontal distance of overhang
$\quad\ H$ = height from glass bottom to overhang
$\quad\ Al$ = noon altitude angle
$\quad\ B$ = roof slope angle (refer to Figure A-2)
$\quad\ C$ = angle formed by intersection of roof slope and altitude (refer to Figure A-2)

The following is an example problem for a sloping overhang. The building is single-story with a large vertical south facade (40° north latitude). Determine the overhang length so that 100 percent of the possible solar radiation will strike the glass on December 21 and 100 percent of

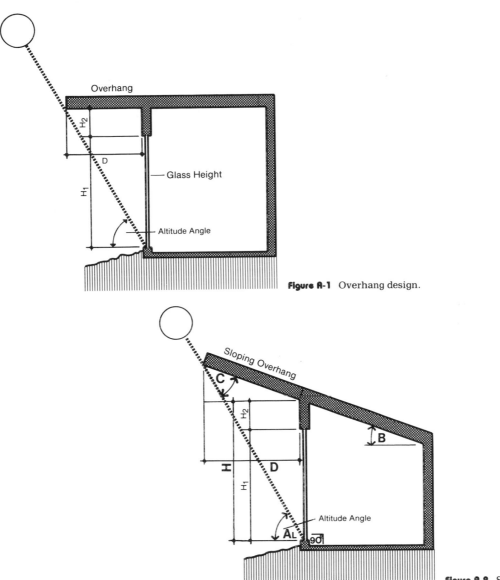

Figure A-1 Overhang design.

Figure A-2 Sloping overhang.

the glass will be shaded at 1 p.m. on May 21 and July 21. The angle of the roof slope is 22.5°; the distance from the top of the glass to the sloping roof is 2 feet; and the glass height is 8 feet. Using the equation, we have

$$C = \text{Al} - B = 48.1°$$

$$D = \frac{(8 \text{ ft} + 2 \text{ ft}) \cos 70.6°}{\sin 48.1°} \cos 22.5°$$

$$= \frac{10 \text{ ft} \times 0.332}{0.744} \times 0.942 = 4.12 \text{ ft}$$

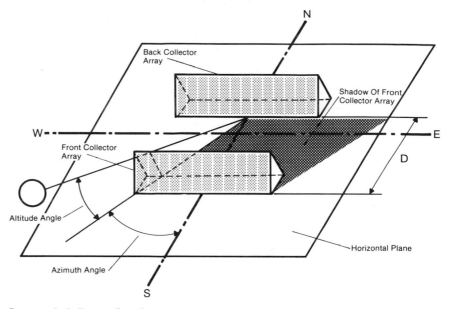

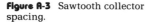

Figure A-3 Sawtooth collector spacing.

Sawtooth Collector Spacing

Often the solar collection area is great enough to demand several arrays or banks of collectors. To ensure solar access to all arrays, they must be spaced far enough apart to eliminate shading. If only midday or noon sun angles are used, the problem is simple. You would simply use the noon altitude angle in the H cot Al equation. However, if shade-free collection during late morning and early afternoon is desirable, the problem is a little more complicated. Assuming that the collector arrays are placed on a flat surface and that they are facing due south, the equation below can be used to determine the distance between two collector arrays. The distance is figured from the front of the first collector array to the front of the second. Figure A-3 is an isometric showing two collector arrays.

$$D = C(\cos B + \sin B \cot Al \cos Az)$$

where D = array spacing in feet
C = collector length in feet
B = collector tilt angle
Al = altitude angle (either morning or afternoon)
Az = azimuth angle (either morning or afternoon)

The following is a sample problem for sawtooth spacing. The active or passive collector arrays are located on a flat surface, facing due south and at 40° north latitude. Determine the spacing between the two arrays at 2 p.m. on December 21. The collector glazing is 10 feet in length, and the collector tilt angle is 55°. Using the equation, we have

$$D = (10 \text{ ft})(\cos 55° + \sin 55° \cot 20.7° \cos 29.4°)$$
$$= (10 \text{ ft})(0.574 + 0.819 \times 2.646 \times 0.871)$$
$$= 24.6 \text{ ft}$$

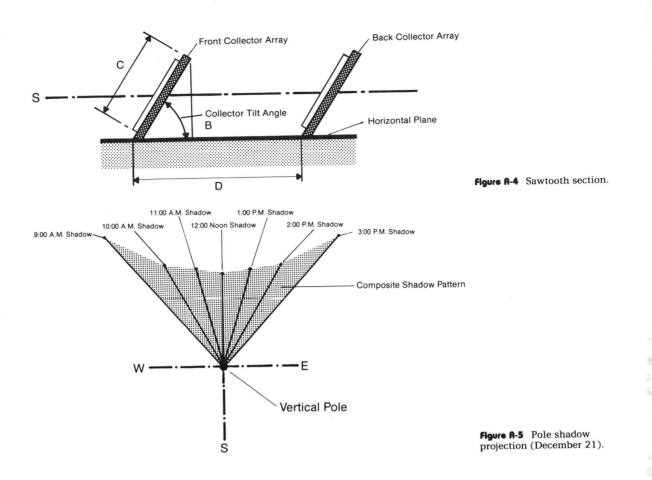

Figure A-4 Sawtooth section.

Figure A-5 Pole shadow projection (December 21).

Building Shadow Patterns

A simple way of determining a building shadow pattern is by using the pole projection method. This method works fairly well for relatively simple building forms. Varying heights around a roof are seen as poles that cast morning, noon, and afternoon shadows. After the pole shadows have been plotted, they can be connected to form a shadow pattern. Figure A-5 illustrates the shadows of a single pole during the hours of sunlight useful for collection on December 21, and Figure A-6 illustrates the shadow pattern of a simple building form. The projected length of each pole can be easily computed by using the equation below. Remember that the shadow lines must follow the altitude angles in the direction of their corresponding azimuth angles.

$$D = H \cot Al \quad \text{or} \quad D = \frac{H}{\tan Al}$$

where D = shadow length in feet at various times
H = building or pole height
Al = altitude angle at various times

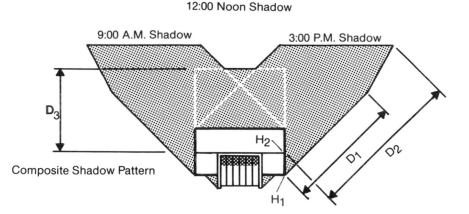

12:00 Noon Shadow

9:00 A.M. Shadow

3:00 P.M. Shadow

D_3

H_2

D_1

D_2

Composite Shadow Pattern

H_1

Figure A-6 Building shadow pattern (December 21).

If the building has a number of roof heights, a number of pole shadows are required to give a proper indication of the shadow pattern. This can be accomplished by hand by using the equation over and over for each height. This may prove to be a tedious process. A simple computer program can be written to reduce the repetition of the required input. All you need to do is input the various heights—H_1, H_2, H_3, etc.—and you get corresponding pole shadow lengths—D_1, D_2, D_3, etc. The shadow pattern can then be constructed by connecting all of the end points of the pole shadows as illustrated in Figure A-6.

For example, construct a shadow pattern for a simple building with a hip roof with the long axis on the east-west axis. The latitude is 48° north latitude. The bottom height of the hip roof is 10 feet, and the ridge is 16 feet from the ground. Using the equation, we have

$$D_1 = H_1 \cot Al \qquad \text{(at 10 a.m. and 2 p.m., Dec. 21)}$$
$$= (10 \text{ ft}) \cot 13.6°$$
$$= 41.3 \text{ ft} \qquad \text{(morning and afternoon shadows)}$$
$$D_2 = H_2 \cot Al \qquad \text{(at 10 a.m. and 2 p.m., Dec. 21)}$$
$$= (16 \text{ ft}) \cot 13.6°$$
$$= 66.1 \text{ ft} \qquad \text{(morning and afternoon shadows)}$$
$$D_3 = H_2 \cot Al \qquad \text{(at noon, Dec. 21)}$$
$$= (16 \text{ ft}) \cot 18.6°$$
$$= 47.5 \text{ ft} \qquad \text{(noon shadow)}$$

Shadow Patterns on Sloping Topography

On sloping topography, shadow lengths will vary. As a consequence, a building shadow pattern may not be symmetrical. Morning and afternoon shadow lengths can vary quite a bit. On uphill slopes the shadows tend to shorten, and on downhill slopes the shadows tend to lengthen. Understanding the effects of topography on shadow patterns can be extremely important for site planning of projects where topography varies from one

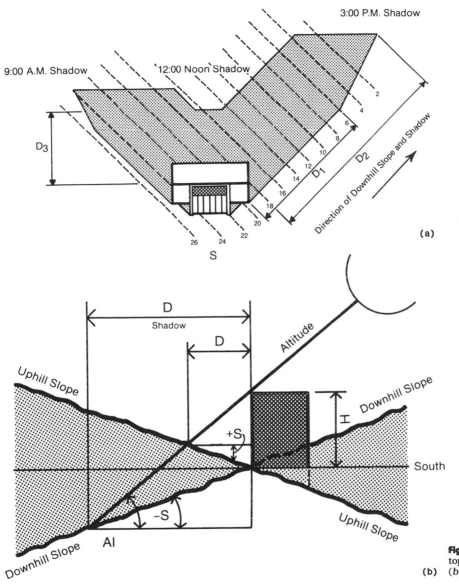

(a)

(b)

building to another. The following equation determines shadow lengths for sloping sites:

$$D = \frac{H}{\tan Al + \tan S}$$

where D = shadow length in feet
H = height of building
Al = altitude angle
S = slope of the site ($+S$ for uphill slopes and $-S$ for downhill slopes)

If we define

$$\tan S = S_a \quad (\text{percent slope}/100)$$

then

$$D = \frac{H}{\tan Al + S_a} \quad (+S_a \text{ for uphill slopes and}$$

$$-S_a \text{ for downhill slopes})$$

For example, determine the morning shadow for a 25-foot-high building with a flat roof at 32° north latitude. The azimuth angle for the shadow is +43.6°, the corresponding altitude angle is 19.8°, and the site slope angle is 3° downhill. Using the equation, we have

$$D = \frac{25 \text{ ft}}{\tan 19.8° + \tan (-3°)}$$

$$= \frac{25 \text{ ft}}{0.36 - 0.052} = 81.2 \text{ ft}$$

The direction of a shadow does not always fall in the direction of a slope. Often the noon shadow will fall in the direction of a slope because of building orientation and the economics of foundation construction. Morning and afternoon shadows, more often than not, will fall off-slope. The following equation augments the previous one and provides for off-slope directions of building shadows.

$$D = \frac{H}{\tan Al + S_a \cos B}$$

where B = angle between the direction of the slope and the direction of the shadow

For example, determine the morning shadow length for a 25-foot-high building with a flat roof located at 32° north latitude. The site slope angle is 3° downhill. The direction of the slope is 15° from the direction of the morning shadow. The azimuth angle for the shadow is +43.6°, and the corresponding altitude angle is 19.8°. Using the equation, we have

$$D = \frac{25 \text{ ft}}{\tan 19.8° + \tan (-3°) \cos 15°}$$

$$= \frac{25 \text{ ft}}{0.36 + (-0.052 \times 0.966)}$$

$$= 80.71 \text{ ft}$$

Computer-Aided Design Techniques

Several computer-aided design techniques have been used throughout this book. With programs developed at the University of Colorado and other universities, building-shading analysis has been made possible. By digitizing the three dimensional coordinates of a building (x, y, and z), a three-dimensional file or record is recorded. From this file many pro-

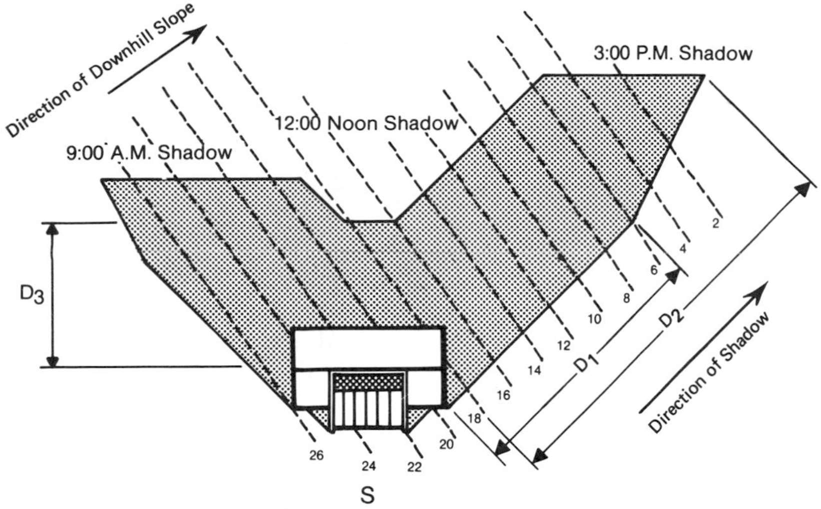

Direction of Downhill Slope
3:00 P.M. Shadow
12:00 Noon Shadow
9:00 A.M. Shadow
D3
D2
D1
Direction of Shadow
2
4
6
8
10
12
14
16
18
20
22
24
26
S

Figure A-8 Off-slope shadow length.

grams can be driven. Brief descriptions of three programs developed by Nelson Greene of the Computer-Aided Design Laboratory (CADLAB) of Boulder, Colorado, follow:

1. **Digit.** Program derived from plans and elevations that file three-dimensional data describing a building or group of buildings.
2. **Eyeball.** Program based upon information filed by the Digit program. This program allows the user to determine any view of the three-dimensional file. A viewing point is determined, and a perspective or isometric is drawn. This program is good if a specific view is desired.
3. **Sunray.** Similar to Eyeball, this program is based upon the sun's-eye view of the object. By selecting a latitude and day of the year, the user can generate perspective drawings for each hour of the day in sunlight. Surfaces in direct view receive direct sunlight. This is an excellent program for analyzing solar access.

Figure A-9 is a series of computer-generated drawings of the Bramwell House located in Florissant, Colorado. Using the Digit and Sunray programs, the perspectives are taken for December 21—the shortest day of the year in the northern hemisphere. This drawing required 800 *x*, *y*, and *z* coordinates to generate. The perspectives represent the sun's-eye view for each of the 9 daylight hours. Note the active collector array at the top of the greenhouse, the vertical south glazing, and the sunspace. All have good exposure throughout the day. There is some shading of the vertical glazings on the first-floor level on either side of the sunspace. This program can be run for June 21, for example, to analyze the effectiveness of the overhangs.

The Sunray program is an excellent analysis tool for many solar-access situations. It may be particularly useful for analyzing self-shadowing caused by building elements, analyzing shading characteristics for projects with multiple buildings, and determining context protection in relation to neighboring buildings, if it is appropriate. The program can be run

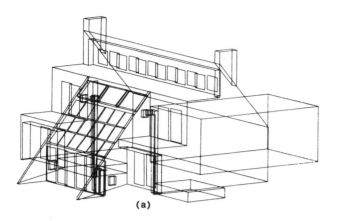

(a)

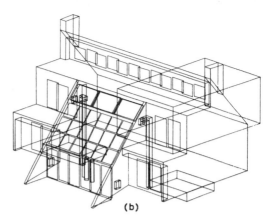

(b)

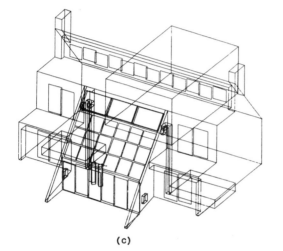

(c)

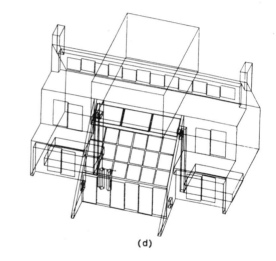

(d)

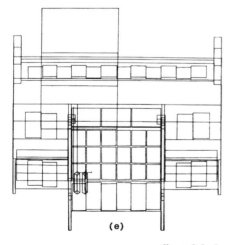

(e)

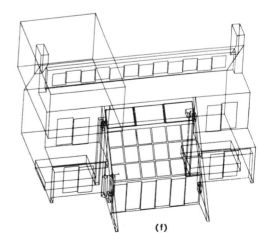

(f)

Figure A-9 Sunray computer analysis: (*a*) 8 a.m., (*b*) 9 a.m., (*c*) 10 a.m., (*d*) 11 a.m., (*e*) 12 noon, (*f*) 1 p.m., (*g*) 2 p.m., (*h*) 3 p.m., and (*i*) 4 p.m. (December 21.)

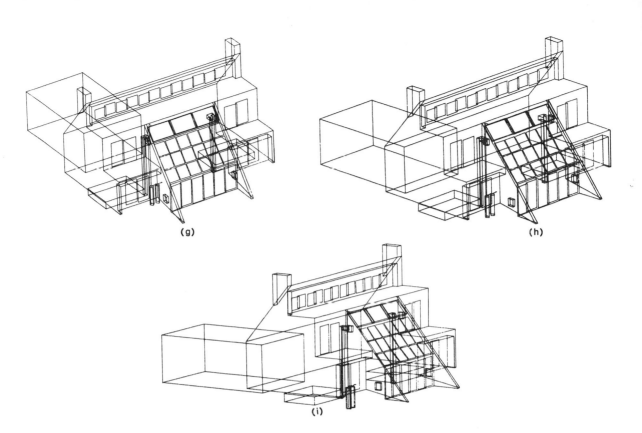

(g)

(h)

(i)

for any day of the year, thereby providing visual information for solar designs for various seasons of the year.

Another useful computer-aided design tool that can be applied to solar-access planning is the computer-drawn shadow mask. This program plots the building profile on the ground for a given day of the year and time of the day. Often several hours of the day can be plotted to create an entire shadow mask. A shading-analysis program was written by student Scott Wolfe of the University of Colorado and was reviewed by Nelson Greene and me. The program plots hourly shadows cast by a single building. Multiple-building shadow plots are possible, but they require more input time. For the purposes of this book, 3 hours of the day were used: 9 a.m., 12 noon, and 3 p.m. Figure A-10 is a computer-generated plot. All of the descriptive elements of the design have been projected in the shadow for each of the 3 hours. Because of all the lines, the shadow is a little complicated. Therefore, in Figure A-11, a drawing enhancing the original computer plot is shown. Both the building and the shadow clearly stand out graphically. For projects with multiple buildings, the building and shadow mask can be reproduced and therefore can be used to group or arrange many buildings on a site plan to provide adequate solar access.

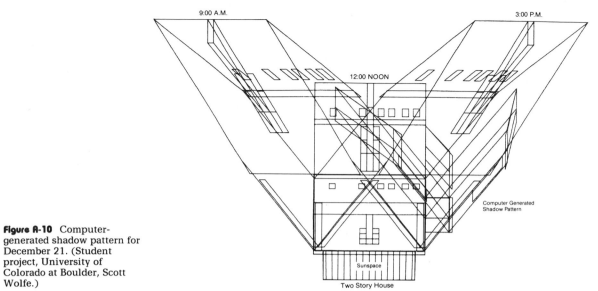

9:00 A.M.   3:00 P.M.

12:00 NOON

Computer Generated
Shadow Pattern

Sunspace

Two Story House

Figure A-10 Computer-generated shadow pattern for December 21. (Student project, University of Colorado at Boulder, Scott Wolfe.)

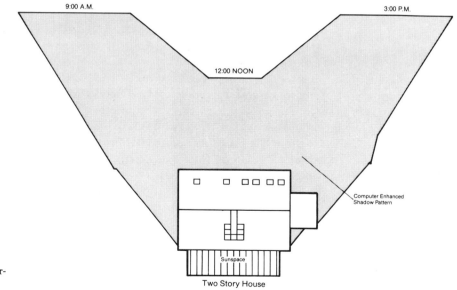

9:00 A.M.   3:00 P.M.

12:00 NOON

Computer Enhanced
Shadow Pattern

Sunspace

Two Story House

Figure A-11 Computer-enhanced drawing.

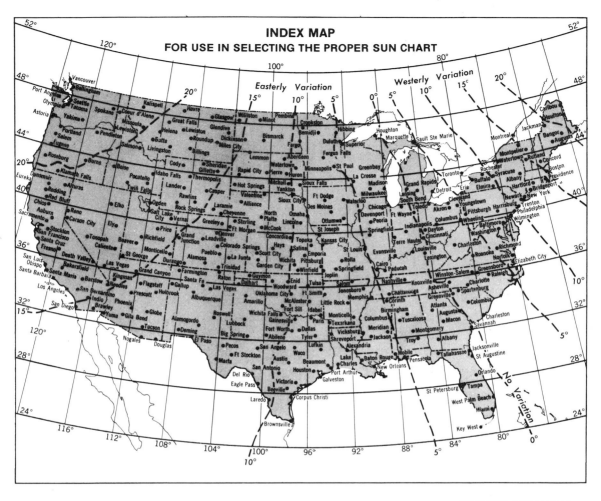

Figure A-12 The isogonic chart of the United States for use in selecting the proper sun chart. (Copyright © 1974 Libby-Owens-Ford Company.) Magnetic variation: The magnetic compass points to magnetic north rather than true north. In most localities magnetic north does not coincide with true north but is toward the east (easterly variation) or toward the west (westerly variation) from it.

The heavy broken lines on this map connect points of equal magnetic variation and present a generalized picture of magnetic variation in the United States. Due to "local attraction," it may be quite different in your locality. For more exact information, consult your local surveyor.

TABLE A-1
Solar Positions and Insolation Values for 24° North Latitude*,²

DATE	SOLAR TIME		SOLAR POSITION		BTUH/SQ. FT. TOTAL INSOLATION ON SURFACES						
	AM	PM	ALT	AZM			SOUTH FACING SURFACE ANGLE WITH HORIZ.				
					NORMAL	HORIZ.	14	*24	34	54	90
JAN 21	7	5	4.8	65.6	71	10	17	21	25	28	31
	8	4	16.9	58.3	239	83	110	126	137	145	127
	9	3	27.9	48.8	288	151	188	207	221	228	176
	10	2	37.2	36.1	308	204	246	268	282	287	207
	11	1	43.6	19.6	317	237	283	306	319	324	226
	12		46.0	0.0	320	249	296	319	332	336	232
	SURFACE DAILY TOTALS				2766	1622	1984	2174	2300	2360	1766
FEB 21	7	5	9.3	74.6	158	35	44	49	53	56	46
	8	4	22.3	67.2	263	116	135	145	150	151	102
	9	3	34.4	57.6	298	187	213	225	230	228	141
	10	2	45.1	44.2	314	241	273	286	291	287	168
	11	1	53.0	25.0	321	276	310	324	328	323	185
	12		56.0	0.0	324	288	323	337	341	335	191
	SURFACE DAILY TOTALS				3036	1998	2276	2396	2446	2424	1476
MAR 21	7	5	13.7	83.8	194	60	63	64	62	59	27
	8	4	27.2	76.8	267	141	150	152	149	142	64
	9	3	40.2	67.9	295	212	226	229	225	214	95
	10	2	52.3	54.8	309	266	285	288	283	270	120
	11	1	61.9	33.4	315	300	322	326	320	305	135
	12		66.0	0.0	317	312	334	339	333	317	140
	SURFACE DAILY TOTALS				3078	2270	2428	2456	2412	2298	1022
APR 21	6	6	4.7	100.6	40	7	5	4	4	3	2
	7	5	18.3	94.9	203	83	77	70	62	51	10
	8	4	32.0	89.0	256	160	157	149	137	122	16
	9	3	45.6	81.9	280	227	227	220	206	186	41
	10	2	59.0	71.8	292	278	282	275	259	237	61
	11	1	71.1	51.6	298	310	316	309	293	269	74
	12		77.6	0.0	299	321	328	321	305	280	79
	SURFACE DAILY TOTALS				3036	2454	2458	2374	2228	2016	488
MAY 21	6	6	8.0	108.4	86	22	15	10	9	9	5
	7	5	21.2	103.2	203	98	85	73	59	44	12
	8	4	34.6	98.5	248	171	159	145	127	106	15
	9	3	48.3	93.6	269	233	224	210	190	165	16
	10	2	62.0	87.7	280	281	275	261	239	211	22
	11	1	75.5	76.9	286	311	307	293	270	240	34
	12		86.0	0.0	288	322	317	304	281	250	37
	SURFACE DAILY TOTALS				3032	2556	2447	2286	2072	1800	246
JUN 21	6	6	9.3	111.6	97	29	20	12	12	11	7
	7	5	22.3	106.8	201	103	87	73	58	41	13
	8	4	35.5	102.6	242	173	158	142	122	99	16
	9	3	49.0	98.7	263	234	221	204	182	155	18
	10	2	62.6	95.0	274	280	269	253	229	199	18
	11	1	76.3	90.8	279	309	300	283	259	227	19
	12		89.4	0.0	281	319	310	294	269	236	22
	SURFACE DAILY TOTALS				2994	2574	2422	2230	1992	1700	204

*Ground reflection not included.

DATE	SOLAR TIME AM	PM	SOLAR POSITION ALT	AZM	BTUH/SQ. FT. TOTAL INSOLATION ON SURFACES NORMAL	HORIZ.	SOUTH FACING SURFACE ANGLE WITH HORIZ. 14	24	34	54	90
JUL 21	6	6	8.2	109.0	81	23	16	11	10	9	6
	7	5	21.4	103.8	195	98	85	73	59	44	13
	8	4	34.8	99.2	239	169	157	143	125	104	16
	9	3	48.4	94.5	261	231	221	207	187	161	18
	10	2	62.1	89.0	272	278	270	256	235	206	21
	11	1	75.7	79.2	278	307	302	287	265	235	32
	12		86.6	0.0	280	317	312	298	275	245	36
	SURFACE DAILY TOTALS				2932	2526	2412	2250	2036	1766	246
AUG 21	6	6	5.0	101.3	35	7	5	4	4	4	2
	7	5	18.5	95.6	186	82	76	69	60	50	11
	8	4	32.2	89.7	241	158	154	146	134	118	16
	9	3	45.9	82.9	265	223	222	214	200	181	39
	10	2	59.3	73.0	278	273	275	268	252	230	58
	11	1	71.6	53.2	284	304	309	301	285	261	71
	12		78.3	0.0	286	315	320	313	296	272	75
	SURFACE DAILY TOTALS				2864	2408	2402	2316	2168	1958	470
SEP 21.	7	5	13.7	83.8	173	57	60	60	59	56	26
	8	4	27.2	76.8	248	136	144	146	143	136	62
	9	3	40.2	67.9	278	205	218	221	217	206	93
	10	2	52.3	54.8	292	258	275	278	273	261	116
	11	1	61.9	33.4	299	291	311	315	309	295	131
	12		66.0	0.0	301	302	323	327	321	306	136
	SURFACE DAILY TOTALS				2878	2194	2342	2366	2322	2212	992
OCT 21	7	5	9.1	74.1	138	32	40	45	48	50	42
	8	4	22.0	66.7	247	111	129	139	144	145	99
	9	3	34.1	57.1	284	180	206	217	223	221	138
	10	2	44.7	43.8	301	234	265	277	282	279	165
	11	1	52.5	24.7	309	268	301	315	319	314	182
	12		55.5	0.0	311	279	314	328	332	327	188
	SURFACE DAILY TOTALS				2868	1928	2198	2314	2364	2346	1442
NOV 21	7	5	4.9	65.8	67	10	16	20	24	27	29
	8	4	17.0	58.4	232	82	108	123	135	142	124
	9	3	28.0	48.9	282	150	186	205	217	224	172
	10	2	37.3	36.3	303	203	244	265	278	283	204
	11	1	43.8	19.7	312	236	280	302	316	320	222
	12		46.2	0.0	315	247	293	315	328	332	228
	SURFACE DAILY TOTALS				2706	1610	1962	2146	2268	2324	1730
DEC 21	7	5	3.2	62.6	30	3	7	9	11	12	14
	8	4	14.9	55.3	225	71	99	116	129	139	130
	9	3	25.5	46.0	281	137	176	198	214	223	184
	10	2	34.3	33.7	304	189	234	258	275	283	217
	11	1	40.4	18.2	314	221	270	295	312	320	236
	12		42.6	0.0	317	232	282	308	325	332	243
	SURFACE DAILY TOTALS				2624	1474	1852	2058	2204	2286	1808

Solar Positions and Insolation Values for 32° North Latitude*²

DATE	SOLAR TIME AM	SOLAR TIME PM	SOLAR POSITION ALT	SOLAR POSITION AZM	BTUH/SQ. FT. TOTAL INSOLATION ON SURFACES NORMAL	HORIZ.	SOUTH FACING SURFACE ANGLE WITH HORIZ. 22	32	42	52	90
JAN 21	7	5	1.4	65.2	1	0	0	0	0	1	1
	8	4	12.5	56.5	203	56	93	106	116	123	115
	9	3	22.5	46.0	269	118	175	193	206	212	181
	10	2	30.6	33.1	295	167	235	256	269	274	221
	11	1	36.1	17.5	306	198	273	295	308	312	245
	12		38.0	0.0	310	209	285	308	321	324	253
	SURFACE DAILY TOTALS				2458	1288	1839	2008	2118	2166	1779
FEB 21	7	5	7.1	73.5	121	22	34	37	40	42	38
	8	4	19.0	64.4	247	95	127	136	140	141	108
	9	3	29.9	53.4	288	161	206	217	222	220	158
	10	2	39.1	39.4	306	212	266	278	283	279	193
	11	1	45.6	21.4	315	244	304	317	321	315	214
	12		48.0	0.0	317	255	316	330	334	328	222
	SURFACE DAILY TOTALS				2872	1724	2188	2300	2345	2322	1644
MAR 21	7	5	12.7	81.9	185	54	60	60	59	56	32
	8	4	25.1	73.0	260	129	146	147	144	137	78
	9	3	36.8	62.1	290	194	222	224	220	209	119
	10	2	47.3	47.5	304	245	280	283	278	265	150
	11	1	55.0	26.8	311	277	317	321	315	300	170
	12		58.0	0.0	313	287	329	333	327	312	177
	SURFACE DAILY TOTALS				3012	2084	2378	2403	2358	2246	1276
APR 21	6	6	6.1	99.9	66	14	9	6	6	5	3
	7	5	18.8	92.2	206	86	78	71	62	51	10
	8	4	31.5	84.0	255	158	156	148	136	120	35
	9	3	43.9	74.2	278	220	225	217	203	183	68
	10	2	55.7	60.3	290	267	279	272	256	234	95
	11	1	65.4	37.5	295	297	313	306	290	265	112
	12		69.6	0.0	297	307	325	318	301	276	118
	SURFACE DAILY TOTALS				3076	2390	2444	2356	2206	1994	764
MAY 21	6	6	10.4	107.2	119	36	21	13	13	12	7
	7	5	22.8	100.1	211	107	88	75	60	44	13
	8	4	35.4	92.9	250	175	159	145	127	105	15
	9	3	48.1	84.7	269	233	223	209	188	163	33
	10	2	60.6	73.3	280	277	273	259	237	208	56
	11	1	72.0	51.9	285	305	305	290	268	237	72
	12		78.0	0.0	286	315	315	301	278	247	77
	SURFACE DAILY TOTALS				3112	2582	2454	2284	2064	1788	469
JUN 21	6	6	12.2	110.2	131	45	26	16	15	14	9
	7	5	24.3	103.4	210	115	91	76	59	41	14
	8	4	36.9	96.8	245	180	159	143	122	99	16
	9	3	49.6	89.4	264	236	221	204	181	153	19
	10	2	62.2	79.7	274	279	268	251	227	197	41
	11	1	74.2	60.9	279	306	299	282	257	224	56
	12		81.5	0.0	280	315	309	292	267	234	60
	SURFACE DAILY TOTALS				3084	2634	2436	2234	1990	1690	370

*Ground reflection not included.

DATE	SOLAR TIME		SOLAR POSITION		BTUH/SQ. FT. TOTAL INSOLATION ON SURFACES						
							SOUTH FACING SURFACE ANGLE WITH HORIZ.				
	AM	PM	ALT	AZM	NORMAL	HORIZ.	22	32	42	52	90
JUL 21	6	6	10.7	107.7	113	37	22	14	13	12	8
	7	5	23.1	100.6	203	107	87	75	60	44	14
	8	4	35.7	93.6	241	174	158	143	125	104	16
	9	3	48.4	85.5	261	231	220	205	185	159	31
	10	2	60.9	74.3	271	274	269	254	232	204	54
	11	1	72.4	53.3	277	302	300	285	262	232	69
	12		78.6	0.0	279	311	310	296	273	242	74
	SURFACE DAILY TOTALS.				3012	2558	2422	2250	2030	1754	458
AUG 21	6	6	6.5	100.5	59	14	9	7	6	6	4
	7	5	19.1	92.8	190	85	77	69	60	50	12
	8	4	31.8	84.7	240	156	152	144	132	116	33
	9	3	44.3	75.0	263	216	220	212	197	178	65
	10	2	56.1	61.3	276	262	272	264	249	226	91
	11	1	66.0	38.4	282	292	305	298	281	257	107
	12		70.3	0.0	284	302	317	309	292	268	113
	SURFACE DAILY TOTALS				2902	2352	2388	2296	2144	1934	736
SEP 21	7	5	12.7	81.9	163	51	56	56	55	52	30
	8	4	25.1	73.0	240	124	140	141	138	131	75
	9	3	36.8	62.1	272	188	213	215	211	201	114
	10	2	47.3	47.5	287	237	270	273	268	255	145
	11	1	55.0	26.8	294	268	306	309	303	289	164
	12		58.0	0.0	296	278	318	321	315	300	171
	SURFACE DAILY TOTALS				2808	2014	2288	2308	2264	2154	1226
OCT 21	7	5	6.8	73.1	99	19	29	32	34	36	32
	8	4	18.7	64.0	229	90	120	128	133	134	104
	9	3	29.5	53.0	273	155	198	208	213	212	153
	10	2	38.7	39.1	293	204	257	269	273	270	188
	11	1	45.1	21.1	302	236	294	307	311	306	209
	12		47.5	0.0	304	247	306	320	324	318	217
	SURFACE DAILY TOTALS				2696	1654	2100	2208	2252	2232	1588
NOV 21	7	5	1.5	65.4	2	0	0	0	1	1	1
	8	4	12.7	56.6	196	55	91	104	113	119	111
	9	3	22.6	46.1	263	118	173	190	202	208	176
	10	2	30.8	33.2	289	166	233	252	265	270	217
	11	1	36.2	17.6	301	197	270	291	303	307	241
	12		38.2	0.0	304	207	282	304	316	320	249
	SURFACE DAILY TOTALS				2406	1280	1816	1980	2084	2130	1742
DEC 21	8	4	10.3	53.8	176	41	77	90	101	108	107
	9	3	19.8	43.6	257	102	161	180	195	204	183
	10	2	27.6	31.2	288	150	221	244	259	267	226
	11	1	32.7	16.4	301	180	258	282	298	305	251
	12		34.6	0.0	304	190	271	295	311	318	259
	SURFACE DAILY TOTALS				2348	1136	1704	1888	2016	2086	1794

TABLE A-3
Solar Positions and Insolation Values for 40° North Latitude*,²

DATE	SOLAR TIME AM	SOLAR TIME PM	SOLAR POSITION ALT	SOLAR POSITION AZM	BTUH/SQ. FT. TOTAL INSOLATION ON SURFACES NORMAL	HORIZ.	SOUTH FACING SURFACE ANGLE WITH HORIZ. 30	40	50	60	90
JAN 21	8	4	8.1	55.3	142	28	65	74	81	85	84
	9	3	16.8	44.0	239	83	155	171	182	187	171
	10	2	23.8	30.9	274	127	218	237	249	254	223
	11	1	28.4	16.0	289	154	257	277	290	293	253
	12		30.0	0.0	294	164	270	291	303	306	263
	SURFACE DAILY TOTALS				2182	948	1660	1810	1906	1944	1726
FEB 21	7	5	4.8	72.7	69	10	19	21	23	24	22
	8	4	15.4	62.2	224	73	114	122	126	127	107
	9	3	25.0	50.2	274	132	195	205	209	208	167
	10	2	32.8	35.9	295	178	256	267	271	267	210
	11	1	38.1	18.9	305	206	293	306	310	304	236
	12		40.0	0.0	308	216	306	319	323	317	245
	SURFACE DAILY TOTALS				2640	1414	2060	2162	2202	2176	1730
MAR 21	7	5	11.4	80.2	171	46	55	55	54	51	35
	8	4	22.5	69.6	250	114	140	141	138	131	89
	9	3	32.8	57.3	282	173	215	217	213	202	138
	10	2	41.6	41.9	297	218	273	276	271	258	176
	11	1	47.7	22.6	305	247	310	313	307	293	200
	12		50.0	0.0	307	257	322	326	320	305	208
	SURFACE DAILY TOTALS				2916	1852	2308	2330	2284	2174	1484
APR 21	6	6	7.4	98.9	89	20	11	8	7	7	4
	7	5	18.9	89.5	206	87	77	70	61	50	12
	8	4	30.3	79.3	252	152	153	145	133	117	53
	9	3	41.3	67.2	274	207	221	213	199	179	93
	10	2	51.2	51.4	286	250	275	267	252	229	126
	11	1	58.7	29.2	292	277	308	301	285	260	147
	12		61.6	0.0	293	287	320	313	296	271	154
	SURFACE DAILY TOTALS				3092	2274	2412	2320	2168	1956	1022
MAY 21	5	7	1.9	114.7	1	0	0	0	0	0	0
	6	6	12.7	105.6	144	49	25	15	14	13	9
	7	5	24.0	96.6	216	214	89	76	60	44	13
	8	4	35.4	87.2	250	175	158	144	125	104	25
	9	3	46.8	76.0	267	227	221	206	186	160	60
	10	2	57.5	60.9	277	267	270	255	233	205	89
	11	1	66.2	37.1	283	293	301	287	264	234	108
	12		70.0	0.0	284	301	312	297	274	243	114
	SURFACE DAILY TOTALS				3160	2552	2442	2264	2040	1760	724
JUN 21	5	7	4.2	117.3	22	4	3	3	2	2	1
	6	6	14.8	108.4	155	60	30	18	17	16	10
	7	5	26.0	99.7	216	123	92	77	59	41	14
	8	4	37.4	90.7	246	182	159	142	121	97	16
	9	3	48.8	80.2	263	233	219	202	179	151	47
	10	2	59.8	65.8	272	272	266	248	224	194	74
	11	1	69.2	41.9	277	296	296	278	253	221	92
	12		73.5	0.0	279	304	306	289	263	230	98
	SURFACE DAILY TOTALS				3180	2648	2434	2224	1974	1670	610

*Ground reflection not included.

DATE	AM	PM	ALT	AZM	NORMAL	HORIZ.	30	40	50	60	90
			SOLAR POSITION		BTUH/SQ. FT. TOTAL INSOLATION ON SURFACES		SOUTH FACING SURFACE ANGLE WITH HORIZ.				
JUL 21	5	7	2.3	115.2	2	0	0	0	0	0	0
	6	6	13.1	106.1	138	50	26	17	15	14	9
	7	5	24.3	97.2	208	114	89	75	60	44	14
	8	4	35.8	87.8	241	174	157	142	124	102	24
	9	3	47.2	76.7	259	225	218	203	182	157	58
	10	2	57.9	61.7	269	265	266	251	229	200	86
	11	1	66.7	37.9	275	290	296	281	258	228	104
	12		70.6	0.0	276	298	307	292	269	238	111
	SURFACE DAILY TOTALS				3062	2534	2409	2230	2006	1728	702
AUG 21	6	6	7.9	99.5	81	21	12	9	8	7	5
	7	5	19.3	90.0	191	87	76	69	60	49	12
	8	4	30.7	79.9	237	150	150	141	129	113	50
	9	3	41.8	67.9	260	205	216	207	193	173	89
	10	2	51.7	52.1	272	246	267	259	244	221	120
	11	1	59.3	29.7	278	273	300	292	276	252	140
	12		62.3	0.0	280	282	311	303	287	262	147
	SURFACE DAILY TOTALS				2916	2244	2354	2258	2104	1894	978
SEP 21	7	5	11.4	80.2	149	43	51	51	49	47	32
	8	4	22.5	69.6	230	109	133	134	131	124	84
	9	3	32.8	57.3	263	167	206	208	203	193	132
	10	2	41.6	41.9	280	211	262	265	260	247	168
	11	1	47.7	22.6	287	239	298	301	295	281	192
	12		50.0	0.0	290	249	310	313	307	292	200
	SURFACE DAILY TOTALS				2708	1788	2210	2228	2182	2074	1416
OCT 21	7	5	4.5	72.3	48	7	14	15	17	17	16
	8	4	15.0	61.9	204	68	106	113	117	118	100
	9	3	24.5	49.8	257	126	185	195	200	198	160
	10	2	32.4	35.6	280	170	245	257	261	257	203
	11	1	37.6	18.7	291	199	283	295	299	294	229
	12		39.5	0.0	294	208	295	308	312	306	238
	SURFACE DAILY TOTALS				2454	1348	1962	2060	2098	2074	1654
NOV 21	8	4	8.2	55.4	136	28	63	72	78	82	81
	9	3	17.0	44.1	232	82	152	167	178	183	167
	10	2	24.0	31.0	268	126	215	233	245	249	219
	11	1	28.6	16.1	283	153	254	273	285	288	248
	12		30.2	0.0	288	163	267	287	298	301	258
	SURFACE DAILY TOTALS				2128	942	1636	1778	1870	1908	1686
DEC 21	8	4	5.5	53.0	89	14	39	45	50	54	56
	9	3	14.0	41.9	217	65	135	152	164	171	163
	10	2	20.7	29.4	261	107	200	221	235	242	221
	11	1	25.0	15.2	280	134	239	262	276	283	252
	12		26.6	0.0	285	143	253	275	290	296	263
	SURFACE DAILY TOTALS				1978	782	1480	1634	1740	1796	1646

TABLE A-4
Solar Positions and Insolation Values for 48° North Latitude*·²

DATE	SOLAR TIME AM	SOLAR TIME PM	SOLAR POSITION ALT	SOLAR POSITION AZM	BTUH/SQ. FT. TOTAL INSOLATION ON SURFACES NORMAL	HORIZ.	SOUTH FACING SURFACE ANGLE WITH HORIZ 38	48	58	68	90
JAN 21	8	4	3.5	54.6	37	4	17	19	21	22	22
	9	3	11.0	42.6	185	46	120	132	140	145	139
	10	2	16.9	29.4	239	83	190	206	216	220	206
	11	1	20.7	15.1	261	107	231	249	260	263	243
		12	22.0	0.0	267	115	245	264	275	278	255
	SURFACE DAILY TOTALS				1710	596	1360	1478	1550	1578	1478
FEB 21	7	5	2.4	72.2	12	1	3	4	4	4	4
	8	4	11.6	60.5	188	49	95	102	105	106	96
	9	3	19.7	47.7	251	100	178	187	191	190	167
	10	2	26.2	33.3	278	139	240	251	255	251	217
	11	1	30.5	17.2	290	165	278	290	294	288	247
		12	32.0	0.0	293	173	291	304	307	301	258
	SURFACE DAILY TOTALS				2330	1080	1880	1972	2024	1978	1720
MAR 21	7	5	10.0	78.7	153	37	49	49	47	45	35
	8	4	19.5	66.8	236	96	131	132	129	122	96
	9	3	28.2	53.4	270	147	205	207	203	193	152
	10	2	35.4	37.8	287	187	263	266	261	248	195
	11	1	40.3	19.8	295	212	300	303	297	283	223
		12	42.0	0.0	298	220	312	315	309	294	232
	SURFACE DAILY TOTALS				2780	1578	2208	2228	2182	2074	1632
APR 21	6	6	8.6	97.8	108	27	13	9	8	7	5
	7	5	18.6	86.7	205	85	76	69	59	48	21
	8	4	28.5	74.9	247	142	149	141	129	113	69
	9	3	37.8	61.2	268	191	216	208	194	174	115
	10	2	45.8	44.6	280	228	268	260	245	223	152
	11	1	51.5	24.0	286	252	301	294	278	254	177
		12	53.6	0.0	288	260	313	305	289	264	185
	SURFACE DAILY TOTALS				3076	2106	2358	2266	2114	1902	1262
MAY 21	5	7	5.2	114.3	41	9	4	4	4	3	2
	6	6	14.7	103.7	162	61	27	16	15	13	10
	7	5	24.6	93.0	219	118	89	75	60	43	13
	8	4	34.7	81.6	248	171	156	142	123	101	45
	9	3	44.3	68.3	264	217	217	202	182	156	86
	10	2	53.0	51.3	274	252	265	251	229	200	120
	11	1	59.5	28.6	279	274	296	281	258	228	141
		12	62.0	0.0	280	281	306	292	269	238	149
	SURFACE DAILY TOTALS				3254	2482	2418	2234	2010	1728	982
JUN 21	5	7	7.9	116.5	77	21	9	9	8	7	5
	6	6	17.2	106.2	172	74	33	19	18	16	12
	7	5	27.0	95.8	220	129	93	77	59	39	15
	8	4	37.1	84.6	246	181	157	140	119	95	35
	9	3	46.9	71.6	261	225	216	198	175	147	74
	10	2	55.8	54.8	269	259	262	244	220	189	105
	11	1	62.7	31.2	274	280	291	273	248	216	126
		12	65.5	0.0	275	287	301	283	258	225	133
	SURFACE DAILY TOTALS				3312	2626	2420	2204	1950	1644	874

*Ground reflection not included.

DATE	SOLAR TIME		SOLAR POSITION		BTUH/SQ. FT. TOTAL INSOLATION ON SURFACES						
	AM	PM	ALT	AZM			SOUTH FACING SURFACE ANGLE WITH HORIZ.				
					NORMAL	HORIZ.	38	48	58	68	90
JUL 21	5	7	5.7	114.7	43	10	5	5	4	4	3
	6	6	15.2	104.1	156	62	28	18	16	15	11
	7	5	25.1	93.5	211	118	89	75	59	42	14
	8	4	35.1	82.1	240	171	154	140	121	99	43
	9	3	44.8	68.8	256	215	214	199	178	153	83
	10	2	53.5	51.9	266	250	261	246	224	195	116
	11	1	60.1	29.0	271	272	291	276	253	223	137
	12		62.6	0.0	272	279	301	286	263	232	144
	SURFACE DAILY TOTALS				3158	2474	2386	2200	1974	1694	956
AUG 21	6	6	9.1	98.3	99	28	14	10	9	8	6
	7	5	19.1	87.2	190	85	75	67	58	47	20
	8	4	29.0	75.4	232	141	145	137	125	109	65
	9	3	38.4	61.8	254	189	210	201	187	168	110
	10	2	46.4	45.1	266	225	260	252	237	214	146
	11	1	52.2	24.3	272	248	293	285	268	244	169
	12		54.3	0.0	274	256	304	296	279	255	177
	SURFACE DAILY TOTALS				2898	2086	2300	2200	2046	1836	1208
SEP 21	7	5	10.0	78.7	131	35	44	44	43	40	31
	8	4	19.5	66.8	215	92	124	124	121	115	90
	9	3	28.2	53.4	251	142	196	197	193	183	143
	10	2	35.4	37.8	269	181	251	254	248	236	185
	11	1	40.3	19.8	278	205	287	289	284	269	212
	12		42.0	0.0	280	213	299	302	296	281	221
	SURFACE DAILY TOTALS				2568	1522	2102	2118	2070	1966	1546
OCT 21	7	5	2.0	71.9	4	0	1	1	1	1	1
	8	4	11.2	60.2	165	44	86	91	95	95	87
	9	3	19.3	47.4	233	94	167	176	180	178	157
	10	2	25.7	33.1	262	133	228	239	242	239	207
	11	1	30.0	17.1	274	157	266	277	281	276	237
	12		31.5	0.0	278	166	279	291	294	288	247
	SURFACE DAILY TOTALS				2154	1022	1774	1860	1890	1866	1626
NOV 21	8	4	3.6	54.7	36	5	17	19	21	22	22
	9	3	11.2	42.7	179	46	117	129	137	141	135
	10	2	17.1	29.5	233	83	186	202	212	215	201
	11	1	20.9	15.1	255	107	227	245	255	258	238
	12		22.2	0.0	261	115	241	259	270	272	250
	SURFACE DAILY TOTALS				1668	596	1336	1448	1518	1544	1442
DEC 21	9	3	8.0	40.9	140	27	87	98	105	110	109
	10	2	13.6	28.2	214	63	164	180	192	197	190
	11	1	17.3	14.4	242	86	207	226	239	244	231
	12		18.6	0.0	250	94	222	241	254	260	244
	SURFACE DAILY TOTALS				1444	446	1136	1250	1326	1364	1304

TABLE A-5
Solar Positions and Insolation Values for 56° North Latitude*·ᶻ

DATE	SOLAR TIME AM	SOLAR TIME PM	SOLAR POSITION ALT	SOLAR POSITION AZM	NORMAL	HORIZ.	46	56	66	76	90
							SOUTH FACING SURFACE ANGLE WITH HORIZ.				
JAN 21	9	3	5.0	41.8	78	11	50	55	59	60	60
	10	2	9.9	28.5	170	39	135	146	154	156	153
	11	1	12.9	14.5	207	58	183	197	206	208	201
	12		14.0	0.0	217	65	198	214	222	225	217
	SURFACE DAILY TOTALS				1126	282	934	1010	1058	1074	1044
FEB 21	8	4	7.6	59.4	129	25	65	69	72	72	69
	9	3	14.2	45.9	214	65	151	159	162	161	151
	10	2	19.4	31.5	250	98	215	225	228	224	208
	11	1	22.8	16.1	266	119	254	265	268	263	243
	12		24.0	0.0	270	126	268	279	282	276	255
	SURFACE DAILY TOTALS				1986	740	1640	1716	1742	1716	1598
MAR 21	7	5	8.3	77.5	128	28	40	40	39	37	32
	8	4	16.2	64.4	215	75	119	120	117	111	97
	9	3	23.3	50.3	253	118	192	193	189	180	154
	10	2	29.0	34.9	272	151	249	251	246	234	205
	11	1	32.7	17.9	282	172	285	288	282	268	236
	12		34.0	0.0	284	179	297	300	294	280	246
	SURFACE DAILY TOTALS				2586	1268	2066	2084	2040	1938	1700
APR 21	5	7	1.4	108.8	0	0	0	0	0	0	0
	6	6	9.6	96.5	122	32	14	9	8	7	6
	7	5	18.0	84.1	201	81	74	66	57	46	29
	8	4	26.1	70.9	239	129	143	135	123	108	82
	9	3	33.6	56.3	260	169	208	200	186	167	133
	10	2	39.9	39.7	272	201	259	251	236	214	174
	11	1	44.1	20.7	278	220	292	284	268	245	200
	12		45.6	0.0	280	227	303	295	279	255	209
	SURFACE DAILY TOTALS				3024	1892	2282	2186	2038	1830	1458
MAY 21	4	8	1.2	125.5	0	0	0	0	0	0	0
	5	7	8.5	113.4	93	25	10	9	8	7	6
	6	6	16.5	101.5	175	71	28	17	15	13	11
	7	5	24.8	89.3	219	119	88	74	58	41	16
	8	4	33.1	76.3	244	163	153	138	119	98	63
	9	3	40.9	61.6	259	201	212	197	176	151	109
	10	2	47.6	44.2	268	231	259	244	222	194	146
	11	1	52.3	23.4	273	249	288	274	251	222	170
	12		54.0	0.0	275	255	299	284	261	231	178
	SURFACE DAILY TOTALS				3340	2374	2374	2188	1962	1682	1218
JUN 21	4	8	4.2	127.2	21	4	2	2	2	2	1
	5	7	11.4	115.3	122	40	14	13	11	10	8
	6	6	19.3	103.6	185	86	34	19	17	15	12
	7	5	27.6	91.7	222	132	92	76	57	38	15
	8	4	35.9	78.8	243	175	154	137	116	92	55
	9	3	43.8	64.1	257	212	211	193	170	143	98
	10	2	50.7	46.4	265	240	255	238	214	184	133
	11	1	55.6	24.9	269	258	284	267	242	210	156
	12		57.5	0.0	271	264	294	276	251	219	164
	SURFACE DAILY TOTALS				3438	2562	2388	2166	1910	1606	1120

*Ground reflection not included.

DATE	SOLAR TIME AM	SOLAR TIME PM	SOLAR POSITION ALT	SOLAR POSITION AZM	BTUH/SQ. FT. TOTAL INSOLATION ON SURFACES NORMAL	HORIZ.	SOUTH FACING SURFACE ANGLE WITH HORIZ 46	56	66	76	90
JUL 21	4	8	1.7	125.8	0	0	0	0	0	0	0
	5	7	9.0	113.7	91	27	11	10	9	8	6
	6	6	17.0	101.9	169	72	30	18	16	14	12
	7	5	25.3	89.7	212	119	88	74	58	41	15
	8	4	33.6	76.7	237	163	151	136	117	96	61
	9	3	41.4	62.0	252	201	208	193	173	147	106
	10	2	48.2	44.6	261	230	254	239	217	189	142
	11	1	52.9	23.7	265	248	283	268	245	216	165
	12		54.6	0.0	267	254	293	278	255	225	173
	SURFACE DAILY TOTALS				3240	2372	2342	2152	1926	1646	1186
AUG 21	5	7	2.0	109.2	1	0	0	0	0	0	0
	6	6	10.2	97.0	112	34	16	11	10	9	7
	7	5	18.5	84.5	187	82	73	65	56	45	28
	8	4	26.7	71.3	225	128	140	131	119	104	78
	9	3	34.3	56.7	246	168	202	193	179	160	126
	10	2	40.5	40.0	258	199	251	242	227	206	166
	11	1	44.8	20.9	264	218	282	274	258	235	191
	12		46.3	0.0	266	225	293	285	269	245	200
	SURFACE DAILY TOTALS				2850	1884	2218	2118	1966	1760	1392
SEP 21	7	5	8.3	77.5	107	25	56	56	54	52	28
	8	4	16.2	64.4	194	72	111	111	108	102	89
	9	3	23.3	50.3	233	114	181	182	178	168	147
	10	2	29.0	34.9	253	146	236	237	232	221	193
	11	1	32.7	17.9	263	166	271	273	267	254	223
	12		34.0	0.0	266	173	283	285	279	265	233
	SURFACE DAILY TOTALS				2368	1220	1950	1962	1918	1820	1594
OCT 21	8	4	7.1	59.1	104	20	53	57	59	59	57
	9	3	13.8	45.7	193	60	138	145	148	147	138
	10	2	19.0	31.3	231	92	201	210	213	210	195
	11	1	22.3	16.0	248	112	240	250	253	248	230
	12		23.5	0.0	253	119	253	263	266	261	241
	SURFACE DAILY TOTALS				1804	688	1516	1586	1612	1588	1480
NOV 21	9	3	5.2	41.9	76	12	49	54	57	59	58
	10	2	10.0	28.5	165	39	132	143	149	152	148
	11	1	13.1	14.5	201	58	179	193	201	203	196
	12		14.2	0.0	211	65	194	209	217	219	211
	SURFACE DAILY TOTALS				1094	284	914	986	1032	1046	1016
DEC 21	9	3	1.9	40.5	5	0	3	4	4	4	4
	10	2	6.6	27.5	113	19	86	95	101	104	103
	11	1	9.5	13.9	166	37	141	154	163	167	164
	12		10.6	0.0	180	43	159	173	182	186	182
	SURFACE DAILY TOTALS				748	156	620	678	716	734	722

SHADOW-LENGTH TABLES _____

Tables A-6 through A-11 give the shadow length for a 1-foot pole on the winter solstice, December 21, for 6 latitudes, 5 slopes, and 3 hours of the day. The morning (a.m.) and afternoon (p.m.) values correspond to azimuth angles of +45° and −45°; the altitude angles also correspond to these azimuth angles. The noon values correspond to a 0° azimuth angle and the zenith angle. These three values define a solar window just under 6 hours for the shortest sunlit day of the year.

To use the tables, first find the latitude table closest to the location of your building site. Second, determine the percent of the slope of the land—from 0 to 20 percent. Third, determine the direction of the slope. The orientations given on the tables represent only downhill slopes. Fourth, going to the table, choose the morning, noon, and afternoon shadow-length values for the 1-foot pole height. Fifth, multiply these values by the actual object or building height to get an actual shadow length.

Example Problems

Find the morning, noon, and afternoon shadow lengths for a two-story building that is 25 feet in height. The building has a flat roof. The building is located at 40° north latitude. The site has a uniform slope of 10 percent downhill to the northeast. Using the tables we find

A.M. value = 4.8 × 25 ft = 120.0 ft
Noon value = 2.3 × 25 ft = 57.5 ft
P.M. value = 9.1 × 25 ft = 227.5 ft

Find the morning, noon, and afternoon shadow lengths for the same building located at 40° north latitude on a site that slopes 10 percent downhill to the southwest. Using the tables we find

A.M. value = 4.8 × 25 ft = 120.0 ft
Noon value = 1.8 × 25 ft = 45.0 ft
P.M. value = 3.2 × 25 ft = 80.0 ft

Note that the afternoon shadow lengths for the two examples vary quite a bit, from 227.5 feet to 80 feet. This is because in the first example the slope is downhill in the direction of the shadow and in the second example the slope is uphill in the direction of the shadow. The figures in the tables are rounded off and therefore may have some errors in shadow lengths for the steeper slopes and taller buildings and at the higher latitudes. These tables were generated by Martin Jaffe and Duncan Erley and published by the American Planning Association in *Protecting Solar Access for Residential Development.*[30a]

TABLE A-6
Shadow Lengths for Latitude 25°30'

Slope	N A.M.	N Noon	N P.M.	NE A.M.	NE Noon	NE P.M.	E A.M.	E Noon	E P.M.	SE A.M.	SE Noon	SE P.M.	S A.M.	S Noon	S P.M.	SW A.M.	SW Noon	SW P.M.	W A.M.	W Noon	W P.M.	NW A.M.	NW Noon	NW P.M.
0%	2.1	1.1	2.1	2.1	1.1	2.1	2.1	1.1	2.1	2.1	1.1	2.1	2.1	1.1	2.1	2.1	1.1	2.1	2.1	1.1	2.1	2.1	1.1	2.1
5%	2.3	1.2	2.3	2.1	1.2	2.4	2.0	1.1	2.3	2.0	1.1	2.1	2.1	1.1	2.0	2.1	1.1	1.9	2.3	1.1	2.0	2.4	1.2	2.1
10%	2.5	1.3	2.5	2.1	1.2	2.7	1.8	1.1	2.5	1.8	1.0	2.1	2.1	1.0	1.8	2.1	1.0	1.7	2.5	1.1	1.8	2.7	1.2	2.1
15%	2.7	1.4	2.7	2.1	1.3	3.1	1.7	1.1	2.7	1.7	1.0	2.1	2.1	1.0	1.7	2.1	1.0	1.6	2.7	1.1	1.7	3.1	1.3	2.1
20%	3.0	1.5	3.0	2.1	1.4	3.6	1.6	1.2	3.0	1.6	1.0	2.1	2.1	0.9	1.6	2.1	1.0	1.5	3.0	1.2	1.6	3.6	1.4	2.1

TABLE A-7
Shadow Lengths for Latitude 30°30'

Slope	N A.M.	N Noon	N P.M.	NE A.M.	NE Noon	NE P.M.	E A.M.	E Noon	E P.M.	SE A.M.	SE Noon	SE P.M.	S A.M.	S Noon	S P.M.	SW A.M.	SW Noon	SW P.M.	W A.M.	W Noon	W P.M.	NW A.M.	NW Noon	NW P.M.
0%	2.7	1.3	2.7	2.7	1.3	2.7	2.7	1.3	2.7	2.7	1.3	2.7	2.7	1.3	2.7	2.7	1.3	2.7	2.7	1.3	2.7	2.7	1.3	2.7
5%	2.7	1.4	2.9	2.7	1.4	3.1	2.4	1.4	2.9	2.4	1.3	2.7	2.4	1.3	2.4	2.7	1.3	2.4	2.9	1.4	2.4	3.1	1.4	2.7
10%	2.7	1.6	3.3	2.7	1.5	3.6	2.2	1.4	3.3	2.2	1.2	2.7	2.2	1.2	2.2	2.7	1.2	2.1	3.3	1.4	2.2	3.6	1.5	2.7
15%	2.7	1.7	3.7	2.7	1.6	4.4	2.1	1.4	3.7	2.1	1.2	2.7	2.1	1.1	2.1	2.7	1.2	1.9	3.7	1.4	2.1	4.4	1.6	2.7
20%	2.7	1.9	4.3	2.7	1.7	5.7	1.9	1.4	4.3	1.9	1.2	2.7	1.9	1.1	1.9	2.7	1.2	1.7	4.3	1.4	1.9	5.7	1.7	2.7

TABLE A-8
Shadow Lengths for Latitude 35°30'

Slope	N A.M.	N Noon	N P.M.	NE A.M.	NE Noon	NE P.M.	E A.M.	E Noon	E P.M.	SE A.M.	SE Noon	SE P.M.	S A.M.	S Noon	S P.M.	SW A.M.	SW Noon	SW P.M.	W A.M.	W Noon	W P.M.	NW A.M.	NW Noon	NW P.M.
0%	3.5	1.6	3.5	3.5	1.6	3.5	3.5	1.6	3.5	3.5	1.6	3.5	3.5	1.6	3.5	3.5	1.6	3.5	3.5	1.6	3.5	3.5	1.6	3.5
5%	4.0	1.8	4.0	3.5	1.7	4.2	3.1	1.6	4.0	3.1	1.5	3.5	3.1	1.5	3.1	3.5	1.5	3.0	4.0	1.6	3.1	4.2	1.7	3.5
10%	4.6	2.0	4.6	3.5	1.8	5.3	2.8	1.6	4.6	2.8	1.5	3.5	2.8	1.4	2.8	3.5	1.5	2.6	4.6	1.6	2.8	5.3	1.8	3.5
15%	5.5	2.2	5.5	3.5	2.0	7.2	2.5	1.6	5.5	2.5	1.4	3.5	2.5	1.3	2.5	3.5	1.4	2.3	5.5	1.6	2.5	7.2	2.0	3.5
20%	6.8	2.5	6.8	3.5	2.2	11.4	2.3	1.7	6.8	2.3	1.3	3.5	2.3	1.3	2.3	3.5	1.3	2.0	6.8	1.7	2.3	11.4	2.2	3.5

TABLE A-9
Shadow Lengths for Latitude 40°30'

Slope	N A.M.	N Noon	N P.M.	NE A.M.	NE Noon	NE P.M.	E A.M.	E Noon	E P.M.	SE A.M.	SE Noon	SE P.M.	S A.M.	S Noon	S P.M.	SW A.M.	SW Noon	SW P.M.	W A.M.	W Noon	W P.M.	NW A.M.	NW Noon	NW P.M.
0%	4.8	2.0	4.8	4.8	2.0	4.8	4.8	2.0	4.8	4.8	2.0	4.8	4.8	2.0	4.8	4.8	2.0	4.8	4.8	2.0	4.8	4.8	2.0	4.8
5%	5.7	2.2	5.7	4.8	2.2	6.2	3.8	2.0	5.7	4.1	1.9	4.8	4.1	1.8	4.1	4.8	1.9	3.8	5.7	2.0	4.1	6.2	2.2	4.8
10%	7.2	2.5	7.2	4.8	2.3	7.2	3.2	2.0	7.2	3.6	1.8	4.8	3.6	1.7	3.6	4.8	1.8	3.2	7.2	2.0	3.6	9.1	2.3	4.8
15%	9.6	2.9	9.6	4.8	2.6	16.6	2.8	2.0	9.1	3.2	1.7	4.8	3.2	1.6	3.2	4.8	1.7	2.8	9.6	2.0	3.2	16.6	2.6	4.8
20%	14.5	3.4	14.5	4.8	2.8	97.5	2.4	2.0	14.5	2.8	1.6	4.8	2.8	1.5	2.8	4.8	1.6	2.4	14.5	2.0	2.8	97.5	2.8	4.8

TABLE A-10
Shadow Lengths for Latitude 45°30'

Slope	N A.M.	N Noon	N P.M.	NE A.M.	NE Noon	NE P.M.	E A.M.	E Noon	E P.M.	SE A.M.	SE Noon	SE P.M.	S A.M.	S Noon	S P.M.	SW A.M.	SW Noon	SW P.M.	W A.M.	W Noon	W P.M.	NW A.M.	NW Noon	NW P.M.
0%	7.2	2.5	7.2	7.2	2.5	7.2	7.2	2.5	7.2	7.2	2.5	7.2	7.2	2.5	7.2	7.2	2.5	7.2	7.2	2.5	7.2	7.2	2.5	7.2
5%	9.6	2.9	9.6	7.2	2.8	11.2	5.7	2.5	9.6	5.3	2.3	7.2	5.7	2.2	5.7	7.2	2.3	5.3	9.6	2.5	5.7	11.2	2.8	7.2
10%	14.6	3.4	14.6	7.2	3.1	25.6	4.8	2.5	14.6	4.2	2.2	7.2	4.8	2.0	4.8	7.2	2.2	4.2	14.6	2.5	4.8	25.6	3.1	7.2
15%	30.3	4.1	30.3	7.2	3.5	—	4.1	2.6	30.3	3.5	2.0	7.2	4.1	1.9	4.1	7.2	2.0	3.5	30.3	2.6	4.1	—	3.5	7.2
20%	—	5.2	—	7.2	4.0	—	3.6	2.6	—	2.9	1.9	7.2	3.6	1.7	3.6	7.2	1.9	2.9	—	2.6	3.6	—	4.0	7.2

TABLE A-11
Shadow Lengths for Latitude 48°30'

Slope	N A.M.	N Noon	N P.M.	NE A.M.	NE Noon	NE P.M.	E A.M.	E Noon	E P.M.	SE A.M.	SE Noon	SE P.M.	S A.M.	S Noon	S P.M.	SW A.M.	SW Noon	SW P.M.	W A.M.	W Noon	W P.M.	NW A.M.	NW Noon	NW P.M.
0%	10.1	3.0	10.1	10.1	3.3	10.1	10.1	3.0	10.1	10.1	3.0	10.1	10.1	3.0	10.1	10.1	3.0	10.1	10.1	3.0	10.1	10.1	3.0	10.1
5%	15.8	3.5	15.8	10.1	3.3	20.5	7.5	3.0	15.8	6.7	2.7	10.1	7.5	2.6	7.5	10.1	2.7	6.7	15.8	3.0	7.5	20.5	3.3	10.1
10%	35.7	4.3	35.7	10.1	3.8	—	5.9	3.0	35.7	5.0	2.5	10.1	5.9	2.3	5.9	10.1	2.5	5.0	35.7	3.0	5.9	—	3.8	10.1
15%	—	5.4	—	10.1	4.4	—	4.9	3.0	—	4.0	2.3	10.1	4.9	2.1	4.9	10.1	2.3	4.0	—	3.0	4.9	—	4.4	10.1
20%	—	7.5	—	10.1	5.2	—	4.2	3.0	—	3.3	2.1	10.1	4.2	1.9	4.2	10.1	2.1	3.3	—	3.0	4.2	—	5.2	10.1

SOLAR COVENANTS B
AND
ORDINANCES

At the present there are several means of providing solar-access protection. Although there is very little legal precedence, many states and communities have adopted some of these methods. There still exist questions concerning the constitutionality of these methods and questions related especially to enforcement of these methods and the exact nature of public benefit to private solar-access protection. Currently there are four methods of providing solar-access protection:

1. **Prior Appropriation.** Prior appropriation is basically a "first come, first serve" method of solar-access protection. It simply provides protection for existing and new buildings, primarily those with more critical solar energy needs.
2. **Private Remedies (Solar Covenants).** This method provides solar-access protection through an agreement between two parties or by a covenant, which may affect several parties. Refer to the sample solar covenant in this appendix.
3. **Public Permits (Solar Ordinances).** Public permits and/or zoning controls can give generalized solar-access protection. Projects must conform to local solar-access ordinances, or they can undergo a special review process in which a permit may be issued.
4. **Public Nuisance Laws.** This method may provide solar-access protection where shading of solar energy is considered to be a public nuisance.

Before examining two of these methods in detail, several issues and questions need to be raised. The practical implementation of solar-access protection is bound to occur with the tremendous growth in installation of solar energy systems in the residential sector especially in more urban contexts. Following are points worth considering before institutionalizing solar-access protection:

1. Does solar-access protection fall under the scope of public health, safety, and welfare; if so, does it justify use of police power?
2. Does one private user have the right to restrict another private user in order to gain solar-access protection?
3. Is the argument of private protection of solar access as a public savings of energy justifiable? Does a specific solar energy system really remove a burden upon public utilities?
4. Is solar-access protection included under the constitutional guarantees of equal protection under the law and due process?

SOLAR-ACCESS COVENANTS

Within the PUD and subdivision processes, solar-access covenants, or solar-rights covenants, can be established. In effect, covenants such as these can provide solar-access protection to individual unit owners. The type of protection may vary from project to project depending upon the level of solar-access protection required and the duration of solar access. When preparing a solar-rights covenant, consider carefully the desired duration of solar access, generate a complete listing of the kinds of solar-access obstructions, and identify the method of enforcing the covenant. The following is a typical solar-access covenant put together using information from the Urban Land Institute's *Residential Development Handbook*[72] and G. Boyer Hayes's *Solar Access Law.*[30]

Example Solar-Access Covenant

No building, part of a building, or structure of any nature, no vegetation or landscaping element, and no other object may be constructed, erected, or planted in such a manner as to intrude or encroach into the sky space, or solar window, designated for solar energy collection. The solar collection device or devices must be free of any shading whatsoever and must be exposed to direct solar radiation (that is, there must be a direct, unobstructed line-of-sight path of the sun to the solar collection device for that period of time commencing at 9 a.m. true solar time and continuing through 3 p.m. true solar time on each day of the year. There shall be no obstruction of or interference with the necessary solar radiation during the time previously specified or with the efficient operation of said solar collection equipment or devices. In the event any building, building part, structure, vegetation, landscaping element, or object of any type is erected or constructed, maintained or cultivated in a manner as to violate the provisions of this covenant, then such violation shall be deemed a public and private nuisance and shall be subject to appropriate action and/or legal proceedings to prevent, enjoin, or remove such nuisance. Legal proceedings to prevent such nuisance may be commenced by the appropriate municipal, county, or state authority or by the owner of the property suffering the aforementioned nuisance.

SOLAR ORDINANCES

Many communities are currently establishing land-use regulations to provide solar-access protection. To make a broad-brush approach to solar-access protection is difficult. The ordinance needs to be open enough not to inhibit development and devalue property. On the other hand, it needs to be tight enough to provide real protection for solar energy systems.

In 1982 the city of Boulder, Colorado, adopted an ordinance to protect the potential for the use of solar energy for home owners and renters while recognizing variations in density and topography. Three solar-access areas were created. The reason for designating three areas was to provide solar-access protection consistent with planned densities, topography, and lot configurations. Descriptions of the three areas follow:

1. **Solar-Access Area I.** This was designated to protect solar access principally for south yards, south walls, and rooftops in districts that currently enjoy such access. Low-density residential districts are included in this area. Zoning districts in area I include RR-E and LR-E.
2. **Solar-Access Area II.** This was designated to protect solar access principally for rooftops in districts where this level of solar access is already enjoyed due to the current level of development. Medium- and high-density residential districts are included in this area. Zoning districts in area II include LR-D, MR-E, MR-D, MR-X, HR-D, HR-D, HR-X, I-E and I-D.
3. **Solar-Access Area III.** This was designated for districts where, because of density, topography, or lot configuration, uniform solar-access protection for south yards, walls, or rooftops may unduly restrict permissible development. Solar-access protection in this area can be obtained through the permit process. Zoning districts in area III include P-E, A-E, CB-E, CB-D, RB-E, RB-D, RB-X, TB-E, and TB-D.

The specific method for determining solar-access protection in these three areas is relatively simple. First of all, the duration of solar-access protection is 4 hours a day. This is especially important on December 21, the shortest day of the year. A solar fence or hypothetical obstruction is the vehicle designed to determine whether or not a building conforms to the solar-access ordinance. In area I a fence height of 12 feet is used; in area II a fence height of 25 feet is used; and in area III there is no fence— in other words, there is no solar-access protection provided through the ordinance. Refer to Figures B-1 and B-2 for illustrations of the solar fence.

A parcel of land can be developed and buildings erected within the bounds of the requirements set by the zoning district, including the buildable area of the site and height limit. A building or structure cannot exceed the specified height limit or a height that would cause the build-

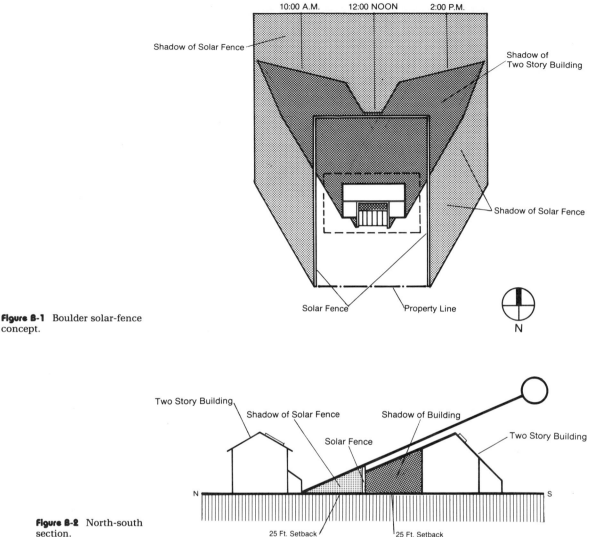

Figure B-1 Boulder solar-fence concept.

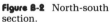

Figure B-2 North-south section.

ing to cast a shadow beyond that of the solar fence. Figure B-1 illustrates a single-family house in area I, where the solar fence is 12 feet in height. The solar fence bounds the east, north, and west sides of the site. The shadow for the fence was determined by 10 a.m., 12 noon, and 2 p.m. sun angles. Concurrently, the 10 a.m., 12 noon, and 2 p.m. shadows of the house are within the bounds of the fence shadows.[55a]

Solar-Access Permits

Solar-access permits may be obtained in order to afford opportunities for solar energy use in circumstances where the basic solar-access ordinance is inadequate. A permit may be obtained by a property owner who intends to construct a solar energy system or by one who already has one in operation. The permit may be obtained through the normal application process. No permit shall restrict use of other property beyond the extent reasonable to ensure efficient and economical beneficial use of solar energy by the permittee.

CONVERSION FACTORS

TABLE C-1
Conversion Factors to SI Units

To convert from	to	Multiply by
	Length	
foot (ft)	meter (m)	3.048×10^{-1}
inch (in)	meter (m)	2.54×10^{-2}
mile (mi)	meter (m)	1.609×10^{3}
	Area	
square foot (ft^2)	square meter (m^2)	9.290×10^{-2}
square inch (in^2)	square meter (m^2)	6.451×10^{-4}
square mile (mi^2)	square meter (m^2)	2.590×10^{6}
	Volume	
cubic foot (ft^3)	cubic meter (m^3)	2.831×10^{-2}
gallon (gal)	cubic meter (m^3)	3.785×10^{-3}
	Velocity (speed)	
feet per second (ft/s)	meters per second (m/s)	3.048×10^{-1}
miles per hour (mi/h or mph)	meters per second (m/s)	4.470×10^{-1}
	Density	
pounds per cubic inch (lb/in^3)	kilograms per cubic meter (kg/m^3)	2.768×10^{4}
grams per cubic centimeter (g/cm^3)	kilograms per cubic meter (kg/m^3)	1.000×10^{3}
	Mass	
pound (lb)	kilogram (kg)	4.536×10^{-1}
gram (g)	kilogram (kg)	1.000×10^{-3}

TABLE C-1 (*Continued*)

To convert from	to	Multiply by
Thermal Conductivity		
Btu-feet per hour per square foot per degree Fahrenheit $[Btu \cdot ft/(h \cdot ft^2 \cdot °F)]$	watts per meter per kelvin [W/ $(m \cdot K)]$	1.731
Force		
pound-force (lb)	newton (N)	4.448
kilogram-force (kg)	newton (N)	9.806
Power		
Btu per hour (Btu/h)	watt (W)	2.931×10^{-1}
foot pound per minute (ft·lb/min)	watt (W)	2.260×10^{-2}
horsepower (hp)	watt (W)	7.457×10^{2}
Energy		
Btu	joule (J)	1.055×10^{3}
calorie (cal)	joule (J)	4.187
foot pound (ft·lb)	joule (J)	1.356
Energy per Unit Area and Time		
Btu per square foot per hour $[Btu/(ft^2 \cdot h)]$	watts per square meter (W/m^2)	3.152

BIBLIOGRAPHY

1. Alexander, C., S. Ishikawa, M. Silverstein, M. Jacobson, I. Fiksdahl, and S. Angel: *A Pattern Language,* Oxford University Press, New York, 1977.

2. American Society of Heating, Refrigerating, and Air-Conditioning Engineers: *ASHRAE Handbook of Fundamentals,* ASHRAE, New York, 1977.

3. Anderson, B., and M. Riordan: *The Solar Home Book,* Cheshire Books, Harrisville, N.H., 1976.

4. Bacon, E.: *Design of Cities,* The Viking Press, New York, 1967.

5. Bainbridge, D., J. Corbett, and J. Hofacre: *Village Homes' Solar House Designs,* Rodale Press, Emmaus, Pa., 1979.

6. Bronowski, J.: *The Ascent of Man,* Little, Brown and Company, Boston, 1973.

7. Brown, G. Z., J. Reynolds, and S. Ubbelohde: *Insideout,* John Wiley & Sons, New York, 1982.

7a. Budyko, M. I.: "The Heat Balance of the Earth's Surface," N. Stepanov, trans., U.S. Department of Commerce, Washington, 1958.

8. Butti, K., and J. Perlin: *A Golden Thread,* Cheshire Books, Palo Alto, Calif., 1980.

9. Caddy, E.: *The Spirit of Findhorn,* Harper & Row, Publishers, New York, 1976.

10. Calder, N.: *Spaceships of the Mind,* British Broadcasting Corporation, London, 1978.

11. Calthorpe, P., and S. Benson: "Beyond Solar: Design for Sustainable Communities," in Gary J. Coates, ed., *Resettling America: Energy, Ecology, and Community,* Brick House Publishing Company, Andover, Mass., 1981.

12. Chermayeff, S., and C. Alexander: *Community and Privacy,* Anchor Books, Doubleday & Company, Garden City, N.Y., 1965.

13. Clark, W.: *Energy for Survival,* Anchor Press, Doubleday & Company, New York, 1974.

14. Clegg, P.: *Energy for the Home,* Garden Way Publishing, Charlotte, Vt., 1975.

15. Corbett, M.: *A Better Place to Live,* Rodale Press, Emmaus, Pa., 1981.

16. Crump, R., and M. Harms: *The Design Connection,* Van Nostrand Reinhold Company, New York, 1981.

17. Daniels, F.: *Direct Use of the Sun's Energy,* Ballantine Books, New York, 1964.

18. Davis, S. (ed.): *The Form of Housing,* Van Nostrand Reinhold Company, New York, 1977.

19. De Chiara, J., and L. Koppelman; *Site Planning Standards,* McGraw-Hill Book Company, New York, 1978.

20. Deudney, D., and C. Flavin: *Renewable Energy: The Power to Choose,* W. W. Norton & Company, New York, 1983.

21. Doczi, G.: *The Power of Limits,* Shambhala Publications, Boulder, Colo., 1981.

22. Doxiadis, C.: *Architecture in Transition,* Oxford University Press, New York, 1963.

23. "Energy Issue," *Progressive Architecture,* April 1980. Copyright 1980, Reinhold Publishing.

24. Fitch, J. M.: "The Aesthetics of Function," in R. Gutman, ed., *People and Buildings,* Basic Books, Inc., Publishers, New York, 1972.

25. Flavin, C.: *Worldwatch Paper 40, Energy and Architecture: The Solar and Conservation Potential,* Worldwatch Institute, November 1982.

26. Flavin, C.: *Worldwatch Paper 52, Electricity from Sunlight: The Future of Photovoltaics,* Worldwatch Institute, December 1982.

27. Fuller, R. B.: *Operating Manual for Spaceship Earth,* E. P. Dutton & Co., New York, 1978.

28. Havlick, S.: *The Urban Organism,* Collier Macmillan Publishers, New York, 1974.

29. Hawkins, G.: *Stonehenge Decoded,* William Collins Sons & Co., Glasgow, 1965.

30. Hayes, G. Boyer: *Solar Access Law,* Ballinger Publishing Co., Cambridge, Mass., 1979.

30a. Jaffe, M., and D. Erley: *Protecting Solar Access for Residential Development: A Guidebook for Planning Officials,* American Planning Association, Chicago, May 1979.

30b. Joint Venture, Inc.: *Proposal to the U.S. Department of Housing and Urban Development,* 1975.

31. Kaysing, W.: *First-Time Farmer's Guide,* Straight Arrow Books, San Francisco, 1971.

32. Knowles, R.: *Sun Rhythm Form,* The MIT Press, Cambridge, Mass., 1981.

33. Kreider, J.: *Medium- and High-Temperature Solar Processes,* Academic Press, New York, 1979.

34. Kreider, J.: *The Solar Heating Design Process,* McGraw-Hill Book Company, New York, 1982.

35. Kreith, F., and J. Kreider: *Principles of Solar Engineering,* McGraw-Hill Book Company, New York, 1978.

36. Landsberg, H. E.: "Climates and Urban Planning in Urban Climates," Tech. Note 108, World Meteorological Organization, Geneva, 1970.

37. Lebens, R.: *Passive Solar Architecture in Europe,* The Architectural Press, Halsted Press Division, London, 1981.

38. Leckie, J., G. Masters, H. Whitehouse, and L. Young: *Other Homes and Garbage,* Sierra Club Books, San Francisco, 1975.

39. Lumpkin, W., and S. Nichols: "Living Proof," *Progressive Architecture,* April 1979.

40. Mazria, E.: *The Passive Solar Energy Book,* Rodale Press, Emmaus, Pa., 1979.

41. McHarg, I.: *Design with Nature,* Doubleday & Company, New York, 1969.

42. Mid-American Solar Energy Complex: *Passive Solar Multi-Family Housing,* Mid-American Solar Energy Complex, 1981.

43. Moffat, A., and M. Schiler: *Landscape Design That Saves Energy,* William Morrow & Company, New York, 1981.

44. Muench, D., and D. Pike: *Anasazi: Ancient People of the Rock,* Crown Publishers, New York, 1974.

45. National Aeronautics and Space Administration: *This Island Earth*, Washington, 1970.

45a. National Association of Home Builders Research Corporation: *Report 1980, Building Survey: Survey of Construction Practices*, NAHBRC, Rockville, Md., 1980.

46. Odum, E. P.: *Ecology: The Link Between the Natural and the Social Sciences*, 2d ed., Holt, Rinehart and Winston, New York, 1975.

47. Olgyay, V.: *Solar Control and Shading Devices*, Princeton University Press, Princeton, N.J., 1957.

48. Olgyay, V.: *Design with Climate*, Princeton University Press, Princeton, N.J., 1963.

49. Organization for Economic Co-operation and Development: "Energy Balances of OCED Countries," OCED, Paris, 1978.

50. Rapoport, A., and N. Watson: "Cultural Variability in Physical Settings," in R. Gutman, ed., *People and Buildings*, Basic Books, Inc., Publishers, New York, 1972.

51. Ridgeway, J., with C. S. Projansky: *Energy-Efficient Community Planning*, The JG Press, The Elements, Emmaus, Pa., n.d.

52. Shurcliff, W. A.: *Air-to-Air Heat Exchangers for Houses*, Wm. A. Shurcliff, Cambridge, Mass., 1981.

53. Simonds, J.: *Earthscape*, McGraw-Hill Book Company, New York, 1978.

54. Smithson, A. (ed.): *Team 10 Primer*, The MIT Press, Cambridge, Mass., 1968.

55. "Solar Architecture," *Process: Architecture*, vol. 6, 1978.

55a. *Solar Access*, Chapter 8 of *Title 9, Land-Use Regulation, Boulder Revised Code*, 1981.

56. Solar Energy Research Institute: *Solar Envelope Zoning: Application to the City Planning Process*, SERI, Golden, Colo., 1980.

57. Solar Energy Research Institute: *Solar Design Workbook*, SERI, Golden, Colo., 1982.

58. Solar Energy Research Institute: *Simplified Housing Energy Design*, SERI, Golden, Colo., unpublished.

59. Stambolis, C.: *Solar Energy in the 80's*, Pergamon Press, Oxford, 1981.

60. Todd, N.: *The Book of the New Alchemists*, E. P. Dutton & Co., New York, 1976.

61. Tunnard, C., and B. Pushkarev: *Man-Made America*, Yale University Press, New Haven, Conn., 1963.

62. Underground Space Center, University of Minnesota: *Earth-Sheltered Community Design*, Van Nostrand Reinhold Company, New York, 1981.

63. Underground Space Center, University of Minnesota: *Earth-Sheltered Homes*, Van Nostrand Reinhold Company, New York, 1981.

64. Underground Space Center, University of Minnesota: *Earth-Sheltered Residential Design Manual*, Van Nostrand Reinhold Company, New York, 1982.

65. U.S. Department of Commerce: *The Climatic Atlas of the United States*, Washington, 1968.

66. U.S. Department of Energy: *Gas Space-Heating Equipment Modifications*, Washington, 1981.

67. U.S. Department of Housing and Urban Development, Office of Policy Development and Research: *Solar Dwelling Design Concepts*, prepared by the American Institute of Architects Research Corporation, Washington, May 1976.

68. U.S. Department of Housing and Urban Development: *The First Passive Solar Home Awards*, Washington, 1979.

69. U.S. Department of Housing and Urban Development: *Protecting Solar Access for Residential Development*, Chicago, 1979.

70. U.S. Department of Housing and Urban Development: *Site Planning for Solar Access*, Chicago, 1980.

71. Untermann, R., and R. Small: *Site Planning for Cluster Housing*, Van Nostrand Reinhold Company, New York, 1977.

72. Urban Land Institute: *Residential Development Handbook*, Urban Land Institute, Washington, 1978.

73. Wadden, R., and P. Scheff: *Indoor Air Pollution,* John Wiley & Sons, New York, 1983. Copyright © 1983 by John Wiley & Sons.

74. Watson, D.: *Designing and Building a Solar House,* Garden Way Publishing, Charlotte, Vt., 1977.

75. Wilhelm, R., trans. (rendered into English by Cary F. Baynes): *The I Ching or Book of Changes,* Bollingen Series 19, Princeton University Press, Princeton, N.J., 1950.

76. Wright, D.: *Natural Solar Architecture,* Van Nostrand Reinhold Company, New York, 1978.

Index

ABOUT THE AUTHOR

Phillip J. Tabb is assistant professor in the College of Design and Planning, the University of Colorado at Boulder, and a principal of Phillip Tabb Architects. He is a member of the American Institute of Architects and the American Planning Association. Over the past 10 years he has designed both passive and active solar systems for a number of large-scale projects, and he was a member of the design team for the Community College of Denver North Campus, one of the largest solar-heated buildings built in the early 1970s. He has won numerous solar research and demonstration grants through the Department of Housing and Urban Development, the Department of Energy, the Solar Energy Research Institute, and the American Institute of Architects Research Corporation. He is also the author of *Here Comes the Sun.*